Agenda Dynamics in Spain

AF698794

Comparative Studies of Political Agendas

Series editors:

Frank R. Baumgartner, Richard J. Richardson Distinguished Professor of Political Science at the University of North Carolina at Chapel Hill, USA

Laura Chaqués-Bonafont, ICREA Professor of Political Science, University of Barcelona and Research Fellow at the Institut Barcelona d'Estudis Internacionals (IBEI), Spain

Christoffer Green-Pedersen, Professor, Aarhus University, Denmark

Arco Timmermans, Research Director, Montesquieu Institute, The Hague and Associate Professor of Comparative Public Policy, Institute of Public Administration, Leiden University, Belgium

Frédéric Varone, Professor of Political Science, University of Geneva, Switzerland

The series publishes books on policy agenda-setting dynamics broadly understood. This includes, for instance, books dealing with the policy effects of agenda dynamics, the relationship between the political agenda, public opinion and the media agenda, and agenda dynamics in relation to particular issues. The series publishes both comparative books and books dealing with single countries, if these single countries are placed in a comparative context. The books can be either monographs or edited volumes.

Titles include:

Laura Chaqués-Bonafont, Anna M. Palau and Frank R. Baumgartner
AGENDA DYNAMICS IN SPAIN

Isabelle Engeli, Christoffer Green-Pedersen and Lars Thorup Larsen (*editors*)
MORALITY POLITICS IN WESTERN EUROPE
Parties, Agendas and Policy Choices

Peter John, Anthony Bertelli, Will Jennings and Shaun Bevan
POLICY AGENDAS IN BRITISH POLITICS

Comparative Studies of Political Agendas Series
Series Standing Order ISBN 978–0–230–35977–2 (Hardback)
978–0–230–35978–9 (Paperback)
(outside North America only)

You can receive future titles in this series as they are published by placing a standing order. Please contact your bookseller or, in case of difficulty, write to us at the address below with your name and address, the title of the series and one of the ISBNs quoted above.

Customer Services Department, Macmillan Distribution Ltd, Houndmills, Basingstoke, Hampshire RG21 6XS, England.

Agenda Dynamics in Spain

Laura Chaqués-Bonafont
ICREA Professor of Political Science, University of Barcelona and IBEI, Spain

Anna M. Palau
Assistant Professor of Political Science, University of Barcelona, Spain

and

Frank R. Baumgartner
Richard J. Richardson Distinguished Professor of Political Science, University of North Carolina at Chapel Hill, USA

palgrave
macmillan

© Laura Chaqués-Bonafont, Anna M. Palau and Frank R. Baumgartner 2015

All rights reserved. No reproduction, copy or transmission of this publication may be made without written permission.

No portion of this publication may be reproduced, copied or transmitted save with written permission or in accordance with the provisions of the Copyright, Designs and Patents Act 1988, or under the terms of any licence permitting limited copying issued by the Copyright Licensing Agency, Saffron House, 6–10 Kirby Street, London EC1N 8TS.

Any person who does any unauthorized act in relation to this publication may be liable to criminal prosecution and civil claims for damages.

The authors have asserted their rights to be identified as the authors of this work in accordance with the Copyright, Designs and Patents Act 1988.

First published 2015 by
PALGRAVE MACMILLAN

Palgrave Macmillan in the UK is an imprint of Macmillan Publishers Limited, registered in England, company number 785998, of Houndmills, Basingstoke, Hampshire RG21 6XS.

Palgrave Macmillan in the US is a division of St Martin's Press LLC, 175 Fifth Avenue, New York, NY 10010.

Palgrave Macmillan is the global academic imprint of the above companies and has companies and representatives throughout the world.

Palgrave® and Macmillan® are registered trademarks in the United States, the United Kingdom, Europe and other countries.

ISBN: 978–1–137–32878–6

This book is printed on paper suitable for recycling and made from fully managed and sustained forest sources. Logging, pulping and manufacturing processes are expected to conform to the environmental regulations of the country of origin.

A catalogue record for this book is available from the British Library.

A catalog record for this book is available from the Library of Congress.

For Fiona Gallach Chaqués

– LC

For Josep Mª Puig Obiols

– AMP

For Jennifer Thompson

– FRB

Contents

List of Figures

List of Tables

Acknowledgments

This book is the result of an ongoing journey that started in 2006, at the joint sessions of the ECPR in Nicosia, Cyprus. At this meeting Bryan Jones, Frank Baumgartner, and John Wilkerson met with Anna M. Palau and Laura Chaqués-Bonafont for the first time, and we collectively discussed the creation of a Spanish agendas project as part of a network then just developing in several European countries following the lead of Christoffer Green-Pedersen in Denmark and based on the US agendas project, which had begun more than ten years earlier. To participate in this international network of scholars meant to be willing to create comprehensive and far-reaching databases, to code thousands of different types of political activities (about 200,000 records of data), and once created, to make the fruit of all this hard work available to anyone for free over an unrestricted website. This was an offer the Spanish team felt they could not refuse. The Americans were talking about the importance of creating and sharing data as a means to do academic research in the field of public policy and comparative politics, and on the Spanish side, we thought this idea was simply awesome and perfectly fit with our academic goals.

The challenge was huge, but motivation and stimulus were greater. For the first two years we devoted most of our time to getting financial support, doing the coding, drafting several papers, and starting to publish the first articles. We received the first grant by the end of 2008 (SEJ-2007-30760-E/SOCI, Ministerio de Ciencia e Innovación), as a part of the EUROCORES collaborative program of the European Science Foundation (ESF) – The Politics of Attention: West European Politics in Times of Change, led by Professor Stefaan Walgrave (University of Antwerp). We received three additional grants from the Spanish government: one from the Ministerio de Ciencia e Innovación (CSO-2009-09397) to analyze the interaction between the political and media agendas; another from the Ministerio de Economia and Competitividad (CSO-2012-31214) to analyze the mobilization of interest groups in the parliamentary and governmental agendas in Spain; and finally, the Ministerio de Ciencia e Innovación also financed the visitor professorship of Professor John Wilkerson to the University of Barcelona in 2008. The collaboration of John Wilkerson has been crucial to develop the project in Spain, and we are extremely grateful for his unconditional support. We had dozens of meetings aimed to adapt the US codebook to the Spanish political context and to discuss about the functioning of political institutions in the US and Spain, which resulted in our first publication, an IBEI working paper in 2008. We maintained our collaboration along these years. In 2009, Luz M. Muñoz Màrquez got a visiting student grant from the Ministerio de Educación to develop some parts of her PhD about Agenda Dynamics and Foreign Aid Policy at the Center of American Politics and Public Policy of the University of Washington, directed by John Wilkerson. Later, Anna M. Palau and Laura Chaqués-Bonafont were also visiting at the Center with a visiting grant of the Institut d'Estudis Autonòmics in 2010.

Other institutions have financed our activities. Our research on morality politics was part of a comparative project about morality politics led by Professor C. Green-Pedersen (Aarhus University, Denmark) and financed to a large extent by the Danish Agency for Science, Technology and Innovation. Also, the Agència de Gestió d'Ajuts Universitaris i de Recerca (AGAUR) granted us with the recognition as an emergent research group from 2009 to 2014, and later as a consolidated research group (2015–), and financed the stay of Professor Baumgartner as a visiting professor at the University of Barcelona in 2012. Actually, our collaboration with Frank Baumgartner started very early in 2007, when he visited us for the first time, and has been growing along these years through countless numbers of meetings discussing the theory and methods of the study of agenda dynamics, mostly in Barcelona but occasionally in Chapel Hill, or during annual meetings of the Comparative Agendas Project or at other conferences. To work with him on the collaboration of this book has been a fascinating and rewarding opportunity, and we are extremely grateful for all his support.

We are grateful to all the institutions that gave us support to develop the project. Also, the Departament de Dret Constitucional i Ciència Política of the Universitat de Barcelona and the Institut Barcelona d'Estudis Internacional, where Laura Chaqués-Bonafont participates as a researcher and member of the Scientific Committee since 2012, have provided the infrastructure and academic support to make this book happen. We are grateful to a large set of scholars who have contributed directly or indirectly to making this book possible. First of all, we want to thank Professor Luz Muñoz Márquez, our friend and colleague in this adventure from the very beginning of the Spanish Policy Agendas project, for her unconditional support and helpful advice in the development of this book. Also, we want to thank the collaboration of the research assistants of the project – Ferran Davesa, Ruben Moliné, Leonor

Flores, Mariel Julio, Lluís Medir, and Andreu Casas – who shared with us countless numbers of meetings about the construction of databases, and also the undergraduate students who participated in coding the media data in 2010: Cristina Capó, Ignasi Torrent, Yair Levy, and Ruben Gordillo.

Since the beginning of the project, we have presented many papers at a number of national and international conferences and workshops, from the ECPR joint sessions and general conferences, to the conferences of the Midwest Political Science Association (MPSA), the American Political Science Association (APSA), the European Political Science Association (EPSA), and the Asociación Española de Ciencia Política (AECPA), among others. We have also participated in the annual meetings of the Comparative Agendas Project, which year after year gather a growing number of scholars from a wider set of countries. In doing this, we have benefited from the helpful comments and suggestions from our colleagues. We are very grateful to all the CAP community, and especially to Christopher Green-Pedersen and Stefaan Walgrave who, besides leading the EUROCORES project of the ESF, edited the first comparative book of the CAP community; to the rest of the editors of Palgrave Macmillan's Comparative Studies of Political Agendas series; Arco Timmermans and Frédéric Varone for their feedback during the project; Peter John and Anthony Bertelli for their provoking comments and ideas during their visit here in Barcelona; Shaun Bevan for his enormous and valuable work; and Marcello Carammia, Emiliano Grossman, Sylvain Brouard, Isabelle Guinaudeau, Gerard Breeman, Will Jennings, Amber Boydstun, and Christian Breuning for their always sharp and interesting comments and suggestions. Jim Stimson was his invaluable, usual selfless person in helping us develop mood measures and generally in understanding the dynamics of public opinion, as only he can do. Last but not least, we would like to thank Professor Bryan Jones, for his invaluable support in this long and fascinating journey. Also, we would like to thank Josep G., Josep M. P., Fiona, Jan and Nora, Núria, Fèlix, Joan, Herminia, Gemma, Natàlia and Albert for their love and unconditional support.

~ *Laura Chaqués-Bonafont and Anna M. Palau*

One of the most fascinating elements of my long involvement in the study of public policy has been the range of collaborations that I have been fortunate enough to develop. As Laura and Anna describe, we began our collaboration almost ten years ago, and it has been a true partnership and a pleasure to get to know them and their team of younger scholars in Barcelona. The book represents one of many kinds of outputs from the Comparative Agendas Project: a theoretically focused exploration of trends and patterns within a single country with the goal of examining that country in detail and in a manner that will make sense to students of that country as well as to those interested and knowledgeable about other Western democracies, and not only about Spain. Were this book written about another country it would of course have different themes, and we would have explored them in different ways. Basque and Catalan nationalism and the role of the Catholic Church would not be the theme of a study of Denmark, and they are far from what Peter John and colleagues (2014) discuss in their companion volume *Policy Agendas in British Politics*. We expect the comparative analysis of policy agendas to include many books like these that focus on a single country, but by no means is this, nor will it be, the predominant approach. Other books, such as that recently edited by Green-Pedersen and Walgrave (2014), will focus on multiple comparisons within countries or, as was covered in the special issue several of us edited in *Comparative Political Studies* (Baumgartner et al., 2011), explicitly comparative approaches of the same issues across many countries with comparable data made available through our multinational collaborations. So the fruit of this work, a book about policy dynamics in Spain, illustrates one of many possible approaches in our multifaceted exploration of public policy issues across Western democracies, which is the Comparative Agendas Project. We and others will do more, and from different angles.

I would like to thank the collaborators mentioned by my colleagues above and to bring particular mention to UNC graduate student Kelsey Shoub, who thought she was going to have a summer job compiling some figures for our book, but who took on much more work than we had bargained for. Juggling course work, exams, teaching responsibilities, and extra work that appear to have no immediate payoff or logic is hard. But it is what we often do as academics, and Kelsey has shown as a young PhD student that she has the capacity to do this, and with grace. It speaks well of her potential for the future. I would also like to express my appreciation to the University of North Carolina at Chapel Hill and for the resources made available to me through the Richard J. Richardson Distinguished Professorship, which funded any number of small needs throughout these years. And, finally, let me express how it has been a joy to collaborate with two knowledgeable, skillful, and insightful professionals such as Laura and Anna. We have made a good team.

~ *Frank R. Baumgartner*

List of Abbreviations

ART	Assisted Reproductive Techniques
BNG	Bloque Nacionalista Gallego (Regional Nationalist Party in Galicia)
CAP	Comparative Agendas Project
CC	Coalición Canaria (Nationalist Party from Canarias)
CCAA	Comunidades Autónomas (Regional Governments)
CDC	Convergència Democràtica de Catalunya (Catalan Nationalist Party making up the CiU coalition)
CHA	Chunta Aragonesista (Aragonese Junta)
CIS	Centro de Investigaciones Sociológicas (Sociological Research Center)
CiU	Convergència i Unió (Catalan Nationalist Party composed of CDC and UDC)
EA	Eusko Alkartasuna (Basque Solidarity)
EAC	European Affairs Committee
ECB	European Central Bank
EEC	European Economic Community
EMU	Economic and Monetary Union
ERC	Esquerra Republicana de Catalunya (left wing Catalan Nationalist Party)
ETA	Euskadi ta Askatasuna (Basque terrorist organization)
EU	European Union
GAL	Grupos Antiterroristas de Liberacion (Anti-terrorist Liberation Group)
GDP	Gross Domestic Product
HB	Herri Batasuna (Basque Homeland and Freedom Party)
IC	Iniciativa per Catalunya (Initiative for Catalonia, Catalan branch of the far left IU)
IU	Izquierda Unida (United Left)
MIP	Adapted series of the Most Important Problem Series compiled by the Centro de Investigaciones Sociológicas
MP	Member of Parliament
NATO	North Atlantic Treaty Organization
PA	Partido Andaluz (Center Left Party from Andalucía)
PAR	Partido Aragonés (Aragonese Party)

PCE	Partido Comunista de España (Spanish Communist Party)
PHN	Plan Hidrológico Nacional (Hidrological National Plan)
PNV	Partido Nacionalista Vasco (Basque Nationalist Party)
PP	Partido Popular (Popular Party)
PSC	Partit Socialista de Catalunya (Catalan Socialist Party)
PSOE	Partido Socialista Obrero Español (Spanish Socialist Worker Party)
SCR	Stem-cell Research
UCD	Unión de Centro Democrático (Center Right Party)
UDC	Unió Democràtica de Catalunya (Catalan Nationalist Party making up the CiU coalition)
UPN	Unión del Pueblo Navarro (Center Right Party, Regionalist Party in Navarra)
WHO	World Health Organization

Part I

Theory and Methods

1
Policy Dynamics in Democratic Spain

Introduction

In this book, we explain how and why policy issues get onto the political agenda. We make both theoretical and empirical contributions to the analysis of policy dynamics and provide a new set of tools for qualitative and quantitative studies of Spanish politics and public policy throughout the period of democratic governance. To do that we follow the policy-dynamics approach. This means that we focus not only on policy preferences and institutional factors, as most comparative analysis does, but also on flows of information, or attention. Our goal is to provide a comprehensive analysis of the patterns of issue prioritization on the agendas of the Spanish government, parliament, and media. We want to go beyond case studies, to cover a time span of more than three decades, and to rely on systematic quantitative indicators for our conclusions. We consider the empirical scope of what we have undertaken to be one of our major contributions to existing research. Of course, we have theoretical and substantive conclusions that go beyond only these methodological points.

Preferences, institutions, and flows of information

Policy preferences are, of course, fundamental to policy dynamics. The government of Socialist Prime Minister José Luis Rodríguez Zapatero put an end to the participation of Spain in the Iraq war, brought marked social change to Spain, promoted women's rights, pushed through contested changes in the laws surrounding abortion, legalized same-sex marriage, and pushed to make the state more secular. Some of these issues were strongly opposed by the conservative Partido Popular (PP), especially those aspects related to defense, international relations, and moral issues such as abortion. By the same token, the priorities defined by PP leader Mariano Rajoy after winning election in November 2011 show important contrasts with previous socialist governments, such as the introduction of a new labor reform, the proposal to reform the regulation of abortion, or the new education law. These are old vindications of the right and were indeed put into practice following its arrival in power.

So, ideological preferences clearly drive many important policy changes (Boix, 1998). The ability of the Spanish state to transform policy preferences into policy outputs is mediated, however, by institutional factors, cultural trends, and unexpected focusing events (Kingdon, 1984; Baumgartner and Jones, 1993, 2015). Institutional factors impose important constraints on the strategies and behaviors of the actors involved in the policymaking process (Hall, 1993). As in other advanced democracies, like the UK, the institutional structure of the Spanish political system generates a bias toward the formation of stable, single-party governments and the domination of the legislative process by the executive (John et al., 2014; Lijphart, 1999; Gunther and Montero, 2009). Most of the time, the governing party does not need to cooperate with opposition parties in order to legislate, which basically implies the prime minister has an almost monopolistic capacity to define the political agenda along the term, especially for those issues whose jurisdiction is not shared across levels of government.

Minority governments have occurred several times. As we explain in detail in this book, minority governments increase the chances for opposition parties to veto the entry of some issues into the agenda, or/and to translate some of their policy priorities into final decisions, like regionalization in the late 1990s. Regional fiscal reform unexpectedly became one of the key priorities of the first government of Aznar. This regulatory change did not respond to the preferences of the PP but to institutional factors. Indeed, this issue was never highlighted during the political campaign of 1996, and in some ways contradicted some of the ideological principles of the PP regarding the territorial organization of power in Spain. Aznar implemented this reform because he was governing with the minority of seats and needed to cooperate with opposition parties, especially regional parties like CiU, to keep a working majority in parliament throughout the rest of his term.

Focusing events and flows of new information are also relevant factors to explain how and why new issues get into the agenda. As the policy-dynamics approach highlights, the policy process is essentially a disorderly and unplanned process dominated in many occasions by unexpected focusing events (Baumgartner and Jones, 1993, 2015). Environmental catastrophes, food-safety scandals, terrorist attacks, corruption scandals related to financial, economic or political institutions, or simply the publication of new ideas and information related to some issues may force a radical change in the issue priorities of the government. The mad-cow outbreak, which appeared to threaten food safety throughout Western Europe; the Al-Qaeda terrorist attacks of September 11, 2001; bankruptcy and misuse of resources of major financial institutions; or the publication of new information concerning the use of citizen's private information by state agencies have had dramatic effects on agenda dynamics, and this should be no surprise.

In this regard, our results indicate that variations in issue attention in the political agenda can be partly predicted by changing economic conditions. A deterioration in economic conditions generates a significant decline in the attention to all issues but the economy and hence, generates a decline in agenda diversity. However, the justifications for dramatic policy reversals also come from other directions. Some are consistent with the ideological tenor of the governing party; some appear to be related to the type of government; some seem to be responses to crisis; and others to changing cultural trends. It is sometimes not the party preferences but the preferences of the prime minister, even if from the same party as the previous prime minister, that explain the ups and downs of some issues in the governmental agenda. By the same token, increasing attention to some issues, especially those related to old political cleavages like political decentralization, terrorism, and rights-related issues, can be predicted by knowing which political parties are governing and which are in the opposition.

In fact, we demonstrate in this book that governing status has a strong impact on the focus of attention or parties in parliament. Political parties, as vote maximizers, select among those issues that may be more rewarding in electoral terms (Petrocik, 1996; Mair 1998; Riker, 1986). While in government, parties tend to focus their attention on a broad set of issues, highlighting those that may give a more positive opinion about what the government is doing, in the media and/or the parliamentary arena. While in the opposition, large-state parties will tend to focus attention on a few sets of issues, especially those which generate more political conflict, leaving aside an important range of issues that perhaps are not so rewarding in electoral terms. In contrast to other countries, in Spain this party conflict is not so much focused on welfare-related issues, but other political cleavages, mainly political decentralization, rights-related issues, and to a lesser extent, terrorism (Fernández-Albertos and Manzano, 2012). These opposition-government dynamics are consistent over time and independent of ideology, contributing significantly to a new politics of confrontation in Spain.

In short, what variations in issue attention have in common is not a single cause, but rather a single process: the political system responds in fits and starts, and in response to many potential causes of change rather than proportionately to the severity of problems or to any single explanatory factor, be it institutional or ideological. While we can find examples of each, an overall theory must focus on the characteristics common to policy change across the board. Neither ideology, preferences, institutional procedures, cultural trends, nor framing by themselves can explain policy change in Spain. Of course, each can and does explain many individual cases. But when we look at policy change in a systematic manner, and across all policy domains, as we do in this book, we find that each explanation falls short.

Multilevel governance and responsiveness

Major institutional transformations, especially the consolidation of a multilevel system of governance, have a direct impact on agenda dynamics. Multilevel governance generates a "double-sided shift of authority away from national governments" (Hooghe and Marks, 2001). In Spain, as in some other countries, this process of delegation of issue jurisdiction upwards to the EU and downwards to the regions happened simultaneously from the 1980s to present, which resulted in a complete transformation of the territorial distribution of power (Hooghe et al., 2010; Morata and Mateo, 2007; Requejo and Miquel, 2010; Subirats and Gallego, 2002). In less than two decades, Spain has been transformed from a unitary state to a highly decentralized state in which the central government only manages half of the public expenditures. In this new context, the policy process is more complex and opaque than before, in terms of the number of public and private actors, the preferences and interest these actors represent, and the pattern of alliances these actors can perform across time and issues in order to accomplish different functions, from exchange of information to the implementation of policy goals (Richardson, 2006; Schmitter, 2004; Peters and Pierce, 2013; Chaqués-Bonafont, 2004).

Multilevel governance means that an increasing share of the political agenda is not defined unilaterally by national political elites. On the one hand, delegation of authority upwards to the EU limits state autonomy to define issue priorities in a wide range of issues, especially those related with the economy, trade, agriculture, and the environment (Brouard et al., 2012; Palau and Chaqués-Bonafont, 2012). On the other hand, regionalization imposes important limits to the agenda-setting capacity of the Spanish government, especially regarding the provision of welfare issues like health or education. To see this, we need look no further than the constitutional reform in 2011, designed to control public deficits according to EU standards, the political confrontation between the Spanish government and some regional governments (especially Catalonia) generated by the new education law passed by the Rajoy government in 2012, or the similar economic policies of governments led by Socialist Prime Minister José Luis Rodriguez Zapatero (2004–2011) and Conservative Mariano Rajoy (2011–2014), both of whom were heavily constrained by the "recommendations" of the EU. While ideological differences are apparent, institutional constraints are also clearly important factors in determining the policy goals and direction of any government.

In this book we provide detailed information about how Europeanization and regionalization affect agenda dynamics. From here, we explain how this transformation has affected mandate responsiveness across time (Stimson et al., 1995; Soroka and Wlezien, 2010; Anderson, 2006). In a multilevel system of governance, the costs to translate policy promises into policy decisions increase because of a greater number of veto-points and higher institutional friction. National governments more and more have to negotiate with a large set of policy actors, which may or may not share their view about what should be done regarding particular issues. At the same time, in a multilevel governance, citizens have a less precise idea about which government is doing what, and thus which government is politically accountable for each policy area. The attribution of responsibilities is more complex and less clearly defined, which may lower the political costs derived from not implementing policy promises and/or not responding to citizens' preferences.

In this context, political parties during electoral campaigns may have strong incentives to keep paying special attention to those issues that are more rewarding in electoral terms, independently of whether they can fulfill their policy promises, or not, once elected. Overall, the governing party will tend to follow its electoral promises (Stimson et al., 1995), but in the case where the party does not succeed it can always start a strategy of blame-shifting oriented to avoid electoral reprisal. In case of inaction, like the so-called *corredor central* (the construction of a high-speed train that connects Spain with France, crossing over the center of the Pyrenees), or in cases where policies clearly conflict with stated ideological positions (as when a left government implements economic adjustment policies in the face of a severe economic recession), political costs can be shifted to other levels of governance. That is, governments can blame the regions or Europe for policies they do not want to embrace but which they implemented. This is facilitated because citizens do not have a full knowledge of which level of government is doing what (Lago and Lago, 2011; Chaqués-Bonafont, Palau and Muñoz 2014), and because, in fact, the situation is more complicated over time.

The incentives of parties to engage in this process of blame-shifting (pushing the blame on other governments when policies fail to respond to citizens' preferences and/or electoral promises) and credit-taking (claiming political responsibility when policies are a success [Anderson, 2006]) are greater for those issues with shared jurisdiction and for those issues that individual parties "own." As the issue ownership theory highlights, parties tend to emphasize those issues where they have a reputation for greater competence and their opponents are less well regarded (Petrocik, 1996). During electoral campaigns, left parties may focus attention and make proposals for policy change about issues related to welfare or the environment where they are perceived to perform better by the public, and they will do so independently of whether they have the jurisdiction to implement these policy proposals. By the same token, right parties may pay more attention and make electoral proposals to issues like the economy, taxes, or the size of government where they have a reputation of good performance, regardless of the increasing role of the EU in these matters. Parties will not hesitate to politicize these issues, even when they know they do not have the jurisdiction to follow through on their policy promises, especially when they can blame others for any eventual failures down the road. As a result, there is a growing imbalance between electoral promises and policy outputs, which contributes to exacerbate the democratic deficit that already characterizes the EU and its member states (Peters and Pierre, 2013; Raunio, 1999; Follesdal and Hix, 2006).

The economic crisis has reinforced this decline in mandate responsiveness. Following the agenda-setting approach, in this book we explain that responsiveness is not only linked to institutional factors but also to external events. Contrary to spatial theories of party behavior (Downs, 1957), the implementation of policy promises does not follow a defined pattern along the period of a government in office, but rather an erratic trend linked to external events (Green-Pedersen and Walgrave, 2014; Chaqués-Bonafont, Palau and Muñoz, 2014; Bevan and Jennings, 2014; John et al., 2011; Bertelli and John, 2013). Governments tend to abandon some of their promises along the term as a response to new conditions (e.g., economic recession) and focusing events (e.g., a terrorist attack), which cannot be predicted during the electoral campaign. As the legislature moves forward, correspondence between promises and policy decisions declines, not necessarily because policy makers are self-interested policy actors or are not interested in fulfilling their electoral promises, but simply because they do not have the cognitive capacity to predict major changes in the political, social, and economic context along the term (Manin, Przeworski and Stokes, 1999; Jones, 2001; Maravall, 2013; Maravall and Sanchez Cuenca, 2008). They must respond, while in government, to the issues that the world throws at them. Often, this means dealing with issues they would prefer to avoid, and which they may not have mentioned during the campaign. So we see important limits to the party mandate theory in the chapters to come, even though we do not contest its motivating logic. Parties may want to implement their programs, but they must govern.

Finally, we test whether mandate responsiveness declines under divided governments (see Klingeman et al., 1994; Budge et al., 2001; Walgrave, Varone and Dumont, 2006; Green-Pedersen, 2006,). As in most other advanced democracies, the capacity to implement policy promises in Spain is greater when the party in government has the majority of seats in the Parliament, and thus does not need to negotiate with opposition parties to pass legislation. However, divided or unitary governments cannot explain why mandate responsiveness declines across time. As argued above, this decline is very much related to the consolidation of multilevel system of governance and unexpected events like economic recession.

Agenda capacity, issue prioritization, and the media

Agenda capacity is among the major factors affecting the ability to transform policy preferences into policy outputs. As the agenda-setting approach illustrates, there is a permanent disequilibrium between the "vast array of issues in need of attention" and the finite size of the agenda, which means that most issues most of the time receive precious little attention (Talbert and Potoski, 2002; Jones and Baumgartner, 2005). In the present context of an economic recession characterized by rising public deficit, deep state fiscal debt, unemployment rates above 25 percent, and nationalist vindications of some regions, there is little room in the Spanish governmental agenda for other issues, despite great urgency. The most important issue is to overcome economic recession, and other important policy reforms – such as the extension of the welfare state or dealing with increasing social and economic inequalities, even where these may be overdue – take a backseat to this more urgent problem. We collectively ignore various important issues because we are overworked dealing with even more urgent ones.

The scarcity of space on the political agenda means that governments sometimes cannot implement even their top priorities. But, over time, the list of urgent problems shifts, and sometimes the agenda opens up for some policy reforms that could not have been made in other periods – not because the reforms were not needed, but simply because other crises were more urgent and left no room for more reforms on the agenda (Baumgartner et al., 2008; Varone et al., 2006; Chaqués-Bonafont and Palau, 2009). Constraints abound on the ability of governments to enact important reforms, but no constraint is as great as the scarcity of attention. Oftentimes, the only reason a given policy reform is not made is that the government, media, and political system are too busy dealing with a more pressing reform. Sometimes, reforms occur at times when, in retrospect, problems were not at their most severe, but when there was space on the agenda because the economy was slack and there was no foreign policy crisis; these are opportunities for domestic policy reforms, but unfortunately, times of simultaneous peace and prosperity are rarer than many Prime Ministers would prefer.

Agenda capacity is constrained by cognitive and institutional factors (Jones and Baumgartner, 2005). Policymakers cannot pay attention to all issues at the same time simply because they do not have the ability to deal with all this information at the same moment. That is, their cognitive capacities are limited (Jones, 2001; Simon, 1947). By the same token, the ability of political elites to introduce new issues into the political debate is constrained by the rules governing the functioning of governmental, parliamentary, and media activities. In some cases, like the Prime Ministers' speeches, there are almost no limits to agenda capacity, meaning the Prime Minister could refer to as many issues as he/she wants while addressing the nation; while in other cases, like oral questions, formal rules impose fixed limits to the number and type of questions per parliamentary group, which obligates MPs to prioritize among issues in each parliamentary question session (Jones et al., 2009; Chaqués-Bonafont and Palau, 2011b).

This book explains how these institutional factors constrain agenda capacity and the implications of this on agenda dynamics. Further, we explain how and why in this process of issue prioritization the media play a key role (Bennett, 1990; Castells, 2009; Walgrave and Van Aelts,

2006). The news media is expected to report to citizens about the major issues of the day, providing the tools for an informed debate between voters, political parties, and the government. The media inform the public about the preferences of political parties and the priorities of the government, and at the same time, constitute a proxy for public opinion, as many politicians rely on the news to understand what the public is or may be concerned about. In doing so, the media should reflect the political and cultural diversity within society, in the sense that all voices can be heard in political deliberations, and there is a fair and impartial balance among policy actors (Norris, 2000). But the media is not a neutral venue for political debate, and most of the time it generates a biased and highly concentrated view of the political world, with attention narrowly focused on just a few high-salience items, leaving great freedom to government officials to operate with virtually no public oversight in other areas.

Our results indicate that in a context of agenda scarcity, the media becomes an instrument for opposition parties to attack governmental political elites (Thesen, 2013). While in the opposition, parties prioritize policy problems, taking into account the issues highlighted in the front pages of newspapers – especially those that are highly controversial – and with greater potential to maximize electoral rewards (Walgrave and Van Aelts, 2006; Green-Pedersen and Stubager, 2010; Baumgartner and Chaqués-Bonafont, 2015). The media became more and more a part of a political process through which political elites interact, blame, and attack each other, emphasizing those issues they consider most important or most likely to be embarrassing to the other side. As a result, the political debate tends to be concentrated in a few issues, and changes in the status quo tend to occur following punctuations.

From the politics of consensus to the politics of conflict

In this book we test some of the key assumptions of the agenda dynamics and political responsiveness theories. In doing so, our goal is also to explain the transformation of Spanish politics from "a model of consensus politics" to a new model characterized by political conflict and increasing confrontation. From the end of Franco's regime after his death in 1975, there developed a consensus about what issues should be prioritized and which decisions were required in order to consolidate the process of democracy (del Aguila, 1982). During the transition to democracy, and later in the 1980s, Spanish politics was mainly about building a new democratic nation in spite of deep political cleavages (mainly state-religion, center-periphery, capital-labor relations) and accommodating adversaries instead of transforming them into enemies. Thus, consensus was the main goal of the executive of Adolfo Suárez when legalizing the Communist party for political parties when drafting the Constitution, for economic elites when signing the *Pactos de la Moncloa,* and for political elites when performing their tasks in and out of the parliamentary arena (Rubio Llorente, 1996). In fact, from 1979 to 1982, almost all the organic laws passed in the Spanish Parliament had the support of the majority of members from the main opposition party (Mújica and Sánchez Cuenca, 2006), which clearly illustrates the willingness of main political elites to agree on certain policy positions about key issues of Spanish politics.

This politics of consensus was the political situation in which some of the most important economic, social, and political transformations have taken place in Spain through the late 1990s (Vallespín, 2012; Gunther and Montero, 2009; Maravall, 2013). Democratization coincided with transformations in the scope and activities of the Spanish state, including marked increases in the size of welfare and regulatory infrastructure as well as a gradual delegation of governmental autonomy upwards to the European Union (EU) and downwards to the regions (Comunidades Autónomas). These transformations – in the functioning of political institutions, the scope of state intervention, and the actors involved in the Spanish democracy – have been so substantial and have occurred so quickly that many of the peculiarities that distinguished Spain from other European democracies had disappeared by the late 1990s (Boix, 2006).

From 1975 to the late 1990s, Spain was transformed from an underdeveloped country (receiving financial support from the OCDE) to a strong economy with a GDP per capita only 10 percent below the average of OCDE countries – with a far-reaching welfare state that manages more than 40 percent of the GDP and shares most of the characteristics that Esping Andersen attributes to the corporatist-statist model institutions – and into one of the countries with better indicators of good governance, despite some structural problems like unemployment, corruption, and terrorism.[1] For more than two decades Spanish political elites have prioritized issues related to the construction of a welfare state, or the consolidation of a market economy, for which there is no significant political conflict (Fernandez-Albertos and Manzano, 2012), or at least not one that is reflected in the set of political parties that contest national elections. On the other hand, this consensual debate skirted important issues, such as the role of the Catholic Church in politics, the politics of memory, or the regulation of moral issues on which the Spanish public and elite institutions are strongly divided.

One of the goals of this book is to explain how, in the new millennium, the politics of resentment has replaced the politics of consensus. Our results illustrate that political parties engage more and more in a strategy of direct confrontation, politicizing highly controversial issues, for which there is an important polarization of voters, and in most of the cases linked with old political cleavages. In some cases, the reactivation of old cleavages is linked to external factors and changing values. The politicization of stem-cell research was almost unavoidable in a context of major scientific advances and technological change, especially in a country like Spain where the medical profession is highly mobilized in favor of the liberalization of this policy (Chaqués-Bonafont and Palau, 2012; Engeli et al., 2012; Varone, 2008). However, in most cases, politicization of issues directly respond to party preferences, like in the case of abortion – or social mobilizations, like the secession vindications in Catalonia – generating the reactivation of old political cleavages.

Political parties do not search for the accommodation of different views and perspectives about what should be done, nor do they search for an open debate in the parliamentary arena in which all political forces participate and reach common policy positions. More and more the political debate is about the politics of scandals, conflicts, and political disputes, that on many occasions are not backed by a coherent discourse or set of ideas about what should be done (Vallespín, 2012: 149). The lack of consensus is also illustrated by a changing governmental style. More than ever, the Spanish Parliament is a political arena that governmental actors simply deny by, among other things, avoiding political debate and governing by decree-law. Because of the partisan nature of governing, this is tolerated by members of the majority parties. But the members of the opposition increasingly respond by voting against governmental proposals (mainly in the case of decree-laws), something that is new in Spanish politics (see Palau, Muñoz and Chaqués-Bonafont, 2014). Our goal is to explain how this confrontation occurs in and out of the parliamentary arena, giving particular attention to the role of the media as a venue for political conflict. To do this, and in contrast to previous research, we rely on a new set of tools that allow for both qualitative and quantitative analysis of policy dynamics over time.

Structure of the book

We have structured the book in nine chapters, organized into four parts. The first part is devoted to theoretical and methodological foundations. In this chapter, we have focused on theoretical expectations, and in Chapter 2, we explain the data and methods used to elaborate our argument and test our hypotheses. In order to make an empirical contribution to the analysis of policy dynamics in Spain, combining quantitative and qualitative analysis, we have created comprehensive and far-reaching databases for the analysis of the political agenda (laws, bills, speeches, oral questions, party manifestos), the media agenda (front-page stories in *El Mundo* and *El País*), and public opinion (a new indicator based on multiple surveys and questions, which we develop following the work of Jim Stimson). All these activities have been coded according to the Comparative Agendas Project (CAP) methodology (Baumgartner et al., 2011). In this second chapter, we also give a general overview of the functioning of the Spanish political system through the lens of our databases (laws, bills, oral questions, speeches, party platforms, media, and public opinion), which provides a starting point for the discussion in the following chapters.

The second part of the book focuses on the dynamics of the political agenda across three institutions of politics and is accordingly organized into three different chapters. Chapter 3 is devoted to the analysis of the governmental agenda. Here our goal is to test whether issue attention varies depending on preferences, institutional factors, and external events. To analyze preferences, we take into consideration party preferences and political leadership and compare among the two; regarding institutional factors, we take as independent variables the type of government and issue jurisdiction across levels of government; and to explore the relevance of external events, we take economic performance as our key explanatory variable. The chapter provides, first, a general description of how issues get in and out of the agenda across time. From here, we test different hypotheses and demonstrate that economic conditions are a much more robust predictor of issue attention than any other indicator (including party ideology, political leadership, changes in issue jurisdiction or institutional supports), that minority governments limit the chances of the executive to pass bills into laws, and that, as in the case of France (Baumgartner et al., 2014), divided governments make no difference for overall legislative productivity in Spain. We also show that the type of government cannot explain why the Spanish government increasingly relies on the decree-law as an instrument to legislate, nor why opposition parties increasingly deny their support to the governing party to pass legislation. Finally, the chapter explains why mandate responsiveness is declining across time.

Chapter 4 explains how the government-opposition game is played in the parliamentary arena. The analysis mainly focuses on variations across parties in issue attention in the oral question time in Parliament. The first part provides detailed information about the functioning and types of oral questions, and especially the differences in agenda capacity across parties. First of all, the fact that large-state parties have an allotment of oral questions ten times larger than small and regional parties explains why issue attention is more concentrated for the CiU, PNV, and/or IU than for the PP or PSOE. Also, we demonstrate that PP and PSOE follow different strategies depending on their institutional status in the government-opposition game, independent of the total number of questions they are allowed to pose each week.

These findings provide a new perspective on the strategic choice of policy topics by political leaders in Spain and make clear the fundamental distinction between governing and opposition parties. More than ideology, governing status determines the focus of attention in the parliamentary arena. Finally, contrary to the logic of the issue-competition approach (Green-Pedersen, 2006), results illustrate that increasing confrontation and polarization have not generated an intensification of parliamentary activities. Actually, the number of oral questions declines from 2008 to present, and this is not linked to economic conditions, or type of government, but rather to changes in parliamentary rules which completely transformed the incentives of the governing party to participate in the oversight sessions. We conclude that these changes have been a fundamental part of the erosion of the politics of consensus that characterized Spanish politics from transition to democracy to the late 1990s and that the Parliament itself has become characterized by increasing confrontation and by an intensification of old political cleavages.

Chapter 5 explains the interrelation between the media and the political agenda. First, contrary to the mediatization theory, results indicate that globalization and changing market dynamics did not generate an increase of so-called soft news. Media coverage of politics today, just as two decades ago, is mainly focused on delivering factual coverage of events and monitoring the activities of political elites. In doing so, the media provides unbalanced coverage of what is going on in the public debate, prioritizing some ideas and framing the stories in ways that respond better to the preferences of their readers. The media is not a neutral venue of politics, and this has important consequences for the functioning of democracy. Results illustrate that media attention is highly concentrated on a very small set of issues. The vast majority of issues get no attention in the front pages of major newspapers and, when they do, attention is mainly focused on a particular set of highly contested debates. We also show that opposition parties tend to follow the media more than governing parties, though this differs dramatically from issue to issue. That is, the media and the main opposition party reinforce each other in this process of issue selection, emphasizing those issues that are more controversial, and thus have a larger capacity for selling newspapers and eroding public support for the governing party.

Chapter 6 explains the politicization of political decentralization. First, we explain the process of delegation of political autonomy to the regions and their capacity for self-rule and shared rule, highlighting the main dimensions of the political debate from the early 1980s to present. Second, we explain changes in public opinion regarding the territorial distribution of power and variation in the prioritization of this issue in the governmental, parliamentary, and media agendas. We argue that political decentralization is increasingly used by political parties to attack each other as a means to maximize electoral rewards. Further, we explain the similarities and differences between the large, state-wide parties and small, regional parties in the politicization of regionalization and demonstrate the importance of media attention in this debate. Political decentralization clearly illustrates the failure of the politics of consensus. It is a central element today of the electoral strategy of most Spanish parties, and hotly contested in elections; this was not previously the case.

The various debates about regionalization in Spain have occurred at the same time as the state has increasingly seen power delegated upwards to the European Union, and this Europeanization of politics is our focus in Chapter 7. To illustrate the dramatic contrasts of the processes of regionalization and Europeanization, we focus on the way political parties frame the delegation of authority upward toward the European Union and downward toward the regions. This means that we analyze not only how these issues have evolved in the political agenda, but especially how political parties have framed these issues, taking into account whether framing strategies are linked to party preferences, public opinion, or changes in economic performance. In this chapter, we show that, contrary to regionalization, the politicization of Europeanization is very low and its saliency has not significantly increased over time; and this is so, despite the fact the EU is increasingly limiting the capacity of Spanish policymakers to define the agenda according to their policy preferences. EU affairs are not among the top priorities of Spanish political parties. Nonetheless, Spanish MPs still use the EU frame when it makes partisan sense to do so and this is particularly linked to the government-opposition game. MPs with governmental status mainly use the EU frame to give visibility to government's achievements and successes. Opposition groups, by contrast, use critical junctures that occur when the proportion of citizens who believe that the country does not benefit from European integration increases to attack the government. Opposition groups also frame questions in EU terms to promote debate on specific issues, especially those that are attractive to their voters and that would be otherwise ignored by the governing party.

Chapter 8 is devoted to morality issues, with a particular focus on abortion. The goal is to trace the different dimensions of the abortion debate in the parliamentary arena and on the front pages of the two most-read newspapers in Spain to analyze the interplay between frames, political parties, and the media. To do so, we develop a list of distinct frames–morality, fairness, efficacy, education, constitutionality, crime and law, and international compatibility – used for or against abortion. We illustrate that political parties emphasize different dimensions of abortion as a problem over time in order to promote or avoid policy change and support their policy preferences. Policy change occurs when there is a shift in the prevailing balance of power, and a new political party wins elections, establishing a new way of thinking about morality issues. Even though the PP was governing with the absolute majority of seats in the Parliament from 2000 to 2004, it did not attempt to reform the existing regulation on abortion toward more conservative positions until 2011 as a response to the new preferences defined by Mariano Rajoy and new Ministry of Justice Alberto Ruiz-Gallardón. On the basis of new moral arguments related to the unacceptable discrimination against a handicapped fetus, the PP has tried to pass a strongly conservative regulatory reform without the support of opposition parties, the medical profession, or an important part of public opinion. The abortion debate is a clear example of the new way of doing politics characterized by increasing confrontation and lack of consensus that we describe over many chapters of this book. Also, by tracing the shifting foci of attention to abortion on the front pages of the two most-read newspapers, we illustrate that the framing strategies of political parties are mirrored in the media in a way consistent with the political parallelism that characterizes the Spanish media system. The framing of abortion in *El País* is closer to the framing strategies of the PSOE and IU, and conservative dimensions of the debate are more highlighted in *El Mundo*, in line with those aspects emphasized by the PP in the parliamentary arena. Overall, the analysis of abortion illustrates that both preferences and institutions are important predictors of agenda access and the role of the media as policy actor that contributes to reinforce tendencies already existing in other areas of policymaking.

Chapter 9 develops our conclusions. We focus particularly on an underdeveloped aspect of politics affecting Spain just as much, but perhaps no more, than other countries: The scarcity of attention. In particular, during periods such as the current one of severe economic crisis, attention is perhaps justly focused on government efforts to improve the economy and to deal with the social issues that follow from the Great Recession. But an underappreciated corollary of this focus of attention is that the government goes on making important decisions affecting the lives of millions of Spanish citizens in many other policy domains, and these are the objects of few official statements and even less media coverage. Our concluding chapter focuses on the lessons of an attention approach to politics and notes the fundamental role of a free and active press in the process, but notes some shortcomings relating to the all-elections focus of so many political leaders and journalists. While it is healthy to focus on electoral competition, the increased focus over time in Spanish politics of what we call the government-opposition game can lead to excessive coverage of a small range of topics, but little attention to thousands of important issues that affect public life, simply because these do not fit into the electoral horse race that often dominates political discourse. Over the period of our study, Spanish politics has become "normalized" as the transition to democracy has become entrenched. At the same time that partisan conflict, rather than regime transformation, has become the defining feature of Spanish politics, so, too, have the negative characteristics of such polarization seeped into the system.

We present a series of appendices as well: Appendix A presents a full list of the policy topics that define the analysis and databases of the Spanish Agendas Project; Appendix B gives a short description of the main political corruption scandals in recent years; Appendix C describes our analysis and coding of events relating to political decentralization; and Appendix D explains our list of EU-related events that form the basis of some of the analysis in Chapter 7.

2
Data and Methods

Having laid out our theoretical approach in the previous chapter, we turn now to the description of the databases used to elaborate our arguments and test our hypotheses. First, we explain how the data was collected and coded to generate the longest known data series developed in Spain for the analysis of agenda dynamics. Second, we give a general overview of these databases using three commonly used indicators in agenda-setting literature – the capacity, diversity, and volatility of the agenda – as a starting point for the discussion in the following chapters. The goal is to assess the level of institutional friction associated with the different parts of the policy process and explain how institutionally imposed costs affect various institutional agendas in Spain.

Data and coding

This book relies on the databases created by the Spanish Policy Agendas Project.[1] These consist of 11 comprehensive, reliable, and comparable data sets that allow tracing the issue priorities of different actors and institutions over a long historical time span. As Table 2.1 illustrates, we have collected information about the agenda of the government (Prime Minister speeches, executive bills, decree-laws), Parliament (legislative-decrees, parliamentary bills, oral questions introduced in plenary meetings and committees), the electoral promises of political parties (party manifestos), the media agenda (front pages of *El Mundo* and *El País*), and public opinion (adapted series of the Most Important Problem Series compiled by the Centro de Investigaciones Sociológicas (CIS)). Although some of these series go back to 1977, most of the databases used in this book cover the period 1982–2013, with the exception of the media and public opinion, which start in 1996 and 1993, respectively.

The data sets were specifically designed to be comprehensive, that is, they are not based on a sample but contain the entire number of bills, legislative acts, and oral questions introduced in the Parliament, and the total number of stories published in the media front pages of the two most read newspapers in Spain. The Prime Minister speeches and party manifestos were also all coded at the quasi-sentence level.[2] Overall, these series add to more than 190,000 records of data that provide a unique set of measures to trace media, public, and policy priorities over time. All the databases were created using publicly available sources (such as the Spanish Parliament webpage),[3] media, and political parties' online archives, and are free and accessible for academics, policy makers, and the general public to download from our webpage.

All the time series have been coded using the methodology of the Comparative Agendas Project (CAP).[4] This consists of 19 major topic categories (23 for the media) (Table 2.1) that are consistent over time and based on policy content. In the media databases, since there are

Table 2.1 Comparative policy agendas topics

1. Economy
2. Rights
3. Health
4. Agriculture
5. Labor
6. Education
7. Environment
8. Energy
10. Transportation
12. Crime & justice
13. Social policy
14. Housing
15. Commerce & banking
16. Defense
17. Science & technology
18. Foreign trade
19. International affairs
20. Government & public administration
21. Public lands & water management
23. Cultural events
27. Weather & natural disasters
29. Sport events
30. Death notices

Notes: A full description of all topics and subtopics is available at www.ub.edu/spanishpolicyagendas. Topics 23, 27, 29 and 30 refer only to media stories, not public policy actions.

many categories of stories, such as sports, weather forecasts or death notices, that are rarely mentioned by policy makers and public opinion, they include few categories that do not exist in the rest of the databases. Because this coding scheme was originally developed to trace change in the policy agenda of the United States, we introduced some revisions in order to capture features that are peculiar of the Spanish system, creating new subtopic categories to capture issues like the social security system, the ETA terrorism, or those related to political decentralization. All in all, our coding system includes 19 topics (23 for the media) broken down into 254 more precise subtopics (see Appendix A). The coding of the data series according to this exhaustive set of topic codes and subcodes allows users of the databases to locate easily all the oral questions, bills or laws introduced in any particular policy domain; for example, water pollution, endangered species protection, unemployment compensation, terrorism, farm subsidies, or military actions abroad. In addition, people can make use of the databases supplementing the series with additional information or by combining or recoding the topic or subtopic categories to meet their research needs.

The topic categories of the CAP project are mutually exclusive and exhaustive, which means that in the inevitable case where a law, oral question or bill covers more than one topic, we code it by the topic that was predominant. Each major topic also includes a general category (numbered as 00) that includes those cases where several different subtopics are mentioned. For example, a law on agriculture covering food safety and animal health would be coded 400 (general agriculture). The complete list of topics and subtopics in the Spanish codebook available on our webpage provides an exhaustive explanation of each, including examples of cases coded into each category (see also Appendix A to this book).

The coding of the data followed a highly systematic and monitored procedure. All the data were coded twice, by coders that were intensively trained to obtain an appropriate understanding of the Spanish codebook. For each series, we calculated reliability scores, counting as errors those cases where both coders did not agree. In the case of the speeches, we obtained a reliability of 93 percent at the topic level and 90 percent at the subtopic level. For the rest of the databases, we used the same method, obtaining equally encouraging scores. We began by coding speeches and continually updated and revised our coding system until solid intercoder reliabilities were achieved. Next, we proceeded to code the rest of the databases, having weekly meetings to solve coding questions. The result is a set of high-quality data sources that, for the first time, can be used to make systematic longitudinal analysis and cross-sectional and cross-country comparisons of policy dynamics. We also adapted the series to the CAP Master codebook that standardizes coding procedures across the 19 countries involved in the project, guaranteeing the comparability of the Spanish databases with those of other CAP teams.

The ability to measure policy attention with this precision allows analyzing aspects of the Spanish political system that have not been subject to systematic empirical study. The substantive agenda[5] of the government and Parliament, for example, is measured in our data series with more than 3000 records of organic laws, ordinary laws, decree-laws, legislative-decrees, and bills. These data provide consistent and reliable information about the asymmetric relationship between the government and Parliament that characterizes Spanish politics from 1978 (Pasquino, 1995; Heywood, 1998; Gunther and Montero, 2009). As we illustrate in Chapter 3, the supremacy of the executive is manifested by the high percentage of decree-laws – they represent 21 percent of the total laws passed from 1982 to 2011 – or by the preeminence of executive bills in front of parliamentary bills – most laws originate in executive bills and almost 90 percent of the bills introduced by parliamentary groups have no chance to go through the parliamentary process.

In this regard, parliamentary bills are more attention-seeking devices than acts designed to generate legislation. They are an important indicator of the symbolic agenda of parliamentary groups, performing a similar function to non-legislative parliamentary activities such as oral questions. As we explain in Chapter 4, oral questions, like parliamentary bills, can be strategically introduced by parliamentary groups to bring some issues to the attention of the Parliament or simply to respond to party preferences or to the demands of the electorate (Wiberg, 1995; Franklin and Norton, 1993; Sanchez de Dios and Wiberg, 2011; Saalfeld, 2000). Our data series include information about oral questions introduced in plenary meetings from 1977 to 2013 (a total of 16,216 questions) and in committees until 2008 (21,076). In addition to information related to the type of question and the name of the parliamentary group and committee, the databases include also a set of specific indicators (see Appendix A) designed to have more detailed information, like the name or gender of the MP that introduces the question, which allow broaching entirely new questions about Spanish politics.

The Prime Minister speeches time series provides an indicator of the symbolic agenda of the executive. The Spanish Prime Minister is more than a *primus inter pares* and has greater powers and resources than his counterparts in most other Western parliamentary democracies (Gunther and Montero, 2009; Vintró, 2006). As Table 2.2 illustrates, since the end of transition to democracy to present (1982–2014), there have been only four Prime Ministers, stability that shows their capacity to effectively control the policy-making process. The Spanish Policy Agendas databases include information about 32 Prime Minister speeches, with nine corresponding to investiture speeches (*Discurso de Investidura*) and 23 to the state-of-the-nation debate (*Debate sobre Política General en torno al Estado de la Nación*). In the investiture speech, the candidate to the presidency of the Spanish government presents his program in order to seek the confidence of the Parliament, providing important information about the policy priorities of the government for the next legislature.[6] In the state-of-the-nation debate, that takes place annually (with the exception of election years), the Prime Minister informs the Parliament about his future policy priorities as well as about the evolution and implementation of the electoral program, informing about variations in the content of the governmental agenda since the beginning of the term. Overall, the speeches databases include a total of 10,049 quasi-sentences, with 2521 corresponding to investiture speeches and 7528 to annual speeches.

Our datasets allow also to systematically analyze how the consolidation of a multilevel governance system has transformed agenda setting and policy making. As we explain in Chapters 6 and 7, public policies in Spain require significant cooperation, as no single level of government is responsible for producing policy outputs. The central government can only prioritize issues within the institutional framework delimited by the Constitution and regional statutes, and consider the policy priorities defined by EU institutions. To capture the complexity that shared jurisdiction introduces in the political system, all the databases include two variables that provide information about the EU and regional content of initiatives.

According to previous research in this topic (Palau and Chaqués-Bonafont, 2012), to identify whether a law, bill, oral question, story in the front pages, or sentence in a speech is linked to the EU, we adopted the following procedure: (1) if a political activity or media story was related to EU treaties and institutions (the Commission, the EU Parliament, etc.), we coded it as 1910; in contrast, (2) when it referred to a specific topic (e.g., food safety), we coded it in the corresponding topic (agriculture is topic 4) and subtopic category (food safety is code 405) and then we added a dummy variable, which was coded 1 if the EU was involved and 0 otherwise. Thus, as in other EU countries participating in the

Comparative Agendas Project, we can track the activities of the national government that have an EU connection, such as laws designed to transpose an EU directive.

In the same vein, following Chaqués-Bonafont (2012), we adopted the following procedure to capture whether media stories and political activities are related to the process of political decentralization. First, all laws, bills, oral questions, speeches, party manifestos or media stories related to the delegation of political autonomy to the regions is coded as 2070. This is a subtopic particular to the Spanish project, designed to capture the dynamics of one of the key issues in Spanish politics, as we explain in detail in Chapter 6.[7] Second, as in the case of Europeanization, all databases include a dummy variable with value 1 for those issues related to regionalization, and 0 otherwise. As an example, all oral questions related to delegation of fiscal policy to the regions is coded as topic 1 (macroeconomics), subtopic 107 (all related to fiscal policy and taxes) and coded 1 for the dummy "regionalization." All political activities related to regionalization are what we call the "regionalization political agenda."

We have also created a dummy variable to identify whether political activities – either bills, oral questions or speeches – are related to self-rule (the capacity of the regions to gain political authority in their territory) and shared rule (the capacity of the regions to be part of the policy-making process at the national level; see appendix C and Hooghe et al., 2008, for further detail). Finally, we also created different databases of the laws passed at regional level in those regions that followed the "fast route" to autonomy provided for in Article 151 of the Spanish Constitution.[8] The regional laws database includes 526 laws passed in Catalonia, 299 in Galicia, 318 in the Basque Country, and 257 in Andalusia.

Institutional factors

Minority governments make decision making more complex mainly because the government cannot pass legislation without the support of third parties. The minority PSOE and PP governments of 1993–1996, 1996–2000, and 2004–2008 formed durable legislative alliances with regional parties (González with CC and CiU until 1995; Aznar with the PNV, CiU and CC; and Zapatero with ERC, IU, BNG, and CHA) to pass legislation (see Table 2.2). The minority government of the PSOE from 2008 to 2011, by contrast, depended on ad hoc agreements with left and regional parties. None of these parties, however, occupied a ministerial position within a coalition government. Interestingly, Felipe González was the only Prime Minister that obtained the support of other parties having the majority of seats from 1982 to 1986.

Table 2.2 Parliamentary legislatures, 1982–2014

Prime Minister	Time in office	Duration in office (months)	Governing party	Votes (%)	Parliamentary seats (%)	Seats of the two main parties (%)	Investiture vote: support of parties
González I	1982–1986	43	PSOE	48	58	88	PCE, CDS, EE
González II	1986–1989	40	PSOE	45	53	83	None
González III	1989–1993	43	PSOE	40	50	81	None
González IV	1993–1996	33	PSOE	39	45	86	CiU, PNV
Aznar I	1996–2000	45	PP	39	45	85	CiU, PNV, CC
Aznar II	2000–2004	47	PP	45	52	88	None
Zapatero I	2004–2008	47	PSOE	43	47	89	ERC, IU, BNG, CHA, CC
Zapatero II	2008–2011	43	PSOE	44	48	92	None
Rajoy	2011–	–	PP	45	53	85	UPN

Note: PSOE (Partido Socialista Obrero Español), PP (Partido Popular), PCE (Partido Comunista de España), CDS (Centro Democrático y Social), EE (Euskadico Esquerra), CiU (Convergència I Unió), PNV (Partido Nacionalista Vasco), CC (Coalición Canaria), CHA (Chunta Aragonesista), ERC (Esquerra republicana de Catalunya), IU (Izquierda Unida), BNG (Bloque Nacionalista Gallego) Votes and Parliamentary seats of the Governing Party.

Source: Electoral data available at *Ministerio del Interior* (www.infoelectoral.mir.es/min/).

As we explain in Chapter 3, the institutional constraints associated with minority governments generally prevent the governing party from fulfilling some of its electoral promises and generate new opportunities for small and regional parties to use their veto power to promote or to limit the entry of some issues into the national agenda (Chaqués-Bonafont and Palau, 2011b, Chaqués-Bonafont, Palau and Muñoz, 2014). The capacity of small and regional parties to be pivotal during minority governments is partly explained by the characteristics of the electoral system that magnify the share of parliamentary seats obtained by the two largest parties – as Table 2.2 illustrates, the PSOE and the PP have always held about 80 percent of the seats in the Spanish Parliament – and from 2004, more than 90 percent – but at the same time, guarantee the representation of small regional parties with concentrated electoral support in their territories (Montero, 1987; Montero and Torcal, 1991; Linz and Montero, 1986; Gunther, Montero and Botella, 2004).

To capture the importance of the type of government on policy dynamics, we added to all the databases a dummy variable that defines as 1 those periods the party in government is governing with the minority of seats, and 0 otherwise. We have also introduced a dummy variable, which identifies whether a political party is giving support to the governing party or not. To calculate this variable, we rely on the voting records of the investiture vote: if the political party voted in favor of the governing party during the investiture vote (e.g., CiU in 1993), we coded that parliamentary group as 1 (see Table 2.2). In most cases, once a political party decides to give support to the governing party, it keeps its agreement for the whole legislature, with some exceptions like CiU at the end of 1995. This decision was very much related to political scandals and corruption linked to the PSOE, and also the fact that in 1995 there were regional elections in Catalonia. These cases of broken agreements are also considered in our databases.

Elections, political promises, and public mood

All databases contain specific variables that capture the importance of the electoral cycle. This allows us to identify whether a bill, oral question, law, speech, or media story takes place during the political campaign (defined here as three months prior to elections), during the election year, or in the last year of a legislature. For the analysis of the electoral promises of political parties, we have coded the party manifesto of the governing party and the main party in opposition. Overall, the data set includes one party manifesto of the UCD (Unión Centro Democrático), 16 of the PSOE, and 16 of the PP. Party manifestos like speeches have been coded at the quasi-sentence level (the data set includes a total of 36,606 quasi-sentences).

Public opinion time series were developed using the data collected by the Centro de Investigaciones Sociológicas (CIS). The CIS reports annual series of citizens' views on the MIP facing the nation from 1993 to present. The MIP series enables us to trace the dynamics of the public agenda over more than 20 years (from 1993 to present), as before 1993 the wording and format of the survey changed dramatically, limiting the possibilities to trace back the series to 1977. MIP answers give accurate information about what problems people think are more important for the nation. To avoid problems related to the variability in question wording, we have coded the answers to the question: "What is, in your opinion, the most important problem that exists today in Spain? And the second? And the third?" from 1993 to 2011. This is an open, multi-answer question that provides information about issue prioritization, but not on citizen's policy positions or preferences for solutions. To translate CIS polls into issue attention percentages we took three steps. First, as the original categorization of the MIP series includes 44 unique categories, excluding "don't know," "no opinion," and "no problem," we coded each answer according to the 19 topics and 247 subtopics of the Policy Agendas Project methodology. For example, "unemployment" answers were coded as subtopic 103 and topic 1. Second, we calculated for each poll the percentages for each issue category; and finally, we aggregated the data on an annual basis, taking the average values in those years where multiple polls were conducted.[9] We also make use of additional public opinion data where consistent series cover substantial periods of time and relate to a public policy of interest. These are used as appropriate in the various chapters to follow and are explained when they are used.

For the analysis of public opinion we also developed an annual and aggregated measure of mood following the methodology developed by Stimson (1991). The mood as originally designed is a measure of the public's preferences for more or less government across multiple policy domains. Stimson developed a complex algorithm for combining as much data as available from public opinion sources in order to develop an index reflecting information contained in multiple sources. His Wcalc algorithm and software is available publicly, and we made use of it to create indices of Spanish public opinion toward the economic and political situation as well as evaluation of national government and EU membership. Wcalc allows one to combine results from multiple opinion sources, provided that each source was measured using the same question wordings and methodologies over time. It assesses the degree to which different question series, which can be from different survey organizations, follow similar trends over time. If they do follow similar trends, then the results may be combined into a single index, making use of more data and therefore being more accurate than any single series could be.

To develop the mood, we used the following survey questions of the CIS for the period 1982–2011:

1. Referring to the economic situation. How do you qualify it: very good, good, normal, bad, very bad;
2. Referring to the political situation. How do you qualify it: very good, good, normal, bad, very bad;
3. Overall, how do you qualify management by the government: very good, good, normal, bad, very bad (available since 1992).

And the following of the Eurobarometer databases:

4. Taking everything into consideration, would you say that (your country) has on balance benefited or not from being a member of the European Community (Common Market)? (available since 1986);
5. Generally speaking, do you think that (your country's) membership of the European Community (Common Market) is a good thing, a bad thing, neither good nor bad (available since 1999).

From here we calculated an "agreement ratio," which is simply the proportion of respondents giving a positive answer divided by those giving either a positive or a negative response (i.e., neutral and "don't know" responses are excluded from the analysis). With these ratios calculated for each survey administration, the Wcalc program generates results similar to a principal components analysis. The algorithm assesses the variation over time within series by rendering them as ratios of the same question repeated over time. When more than one question series is used for a policy-specific mood, it assesses covariation between series by observing the covariation of those ratios over time. Series that do not correspond are not included in the index (though, in our case, all of the series mentioned above do load highly on the index and are maintained).

The dyad ratios algorithm assesses first whether each component question series loads highly with the other question series. If, for example, a given survey organization or question wording generates responses consistently lower by, say, five points, than another series, but both series show similar trends over time, then the algorithm will show that they correspond to a similar index because of their shared variation over time. The method has been used in many cases including studying public opinion toward the death penalty, where Baumgartner and colleagues (2008) combined multiple different questions on the topic to generate a highly robust index; by Brouard and others (2014) looking at French opinion toward nuclear power; and by Stimson, Tiberj, and others (2012, 2013) concerning French public mood in general.

Media data

Finally, for the study of the media agenda we developed a comprehensive and far-reaching database of the stories covered in the front pages of the two most relevant Spanish newspapers in Spain from 1996 to 2011. The analysis of how media attention gets distributed across issues is interesting, in order to understand the role of the media as political actor, but also because the news strongly influences citizens and policy makers' responses to policy issues (e.g., Boydstun, 2013; Wolfe et al., 2013; Chaqués-Bonafont and Baumgartner, 2013; Baumgartner and Chaqués-Bonafont, 2014; Walgrave et al., 2007). Our media data series include 107,308 stories, with 56,541 for *El País* and 50,767 for *El Mundo*. As for the rest of the databases, this is a comprehensive data set (not a sample) that includes information about all the stories published in the front pages of these two newspapers. We focused on front-page stories because they reflect the issues that editors and journalists consider the most important or striking and are also the first thing that people see when reading a newspaper, capturing most of the attention of the public in general. The media data series include also a set of specific indicators (see Appendix A) designed to gather information about a broad range of issues like whether the stories are domestic or international, if they have political content, or the type of actors mentioned, from political parties to interest groups.

Overall, all the chapters in this book have in common the use of data series developed by the Spanish Policy Agendas Project that generate the longest-known data series about the issue preferences of different actors and institutions in Spain. Next, we give a general description of these databases, considering to what extent variations in the institutional friction associated with different policy-making channels affect agenda capacity, diversity, and volatility.

Agenda capacity

Spanish citizens consider that many issues are worthy of policy makers' attention. Most believe that the government should put an end to ETA terrorism, reduce school failure, or cut down the high unemployment rates that have historically characterized the Spanish economy. Some issues, like terrorism or unemployment, have consistently been part of the Spanish political agenda from the end of Francoism to the present, while others have entered the agenda more recently. The development of a modern welfare state or the increasing complexity and secularization of Spanish society has led new issues, like housing, euthanasia, or same-sex marriage, to be considered among some people also as issues worthy of political attention. However, individual policy makers and institutions have limited capacity to be involved simultaneously in multiple issues. They lack the capacity to do everything that is asked of them because of limitations in terms of time, energy, resources, or cognitive capacity (Simon, 1971; Baumgartner and Jones, 2002; Talbert and Potoski, 2002; Jones and Baumgartner, 2005).

Agenda capacity, namely the information that an institution or individual policy maker can process in a specific period of time, depends on cognitive constraints (related to the limited processing capacity of human beings), but also on the amount of institutional friction imposed by different policy activities (Jones and Baumgartner, 2005). Institutional friction refers to the costs imposed by particular rules and procedures and is mainly associated with *decision costs* (those that actors incur when trying to come to an agreement) and *transaction costs* (associated with enforcement and compliance after a decision is taken). Contrary to other types of costs, like those associated with obtaining relevant information to take a decision (*information costs*), decision and transaction costs are highly sensitive to particular institutional rules. Further, these decision-making costs increase as we move through the policy cycle (Chaqués-Bonafont and Palau 2011b). Early stages of agenda setting, like speeches, are subjected to less institutional friction than later stages of decision making, like organic laws or bills, which are associated with higher decision costs.

In Table 2.3, we summarize and rank the different policy-making channels considered in this book in terms of institutional friction, from lowest to highest levels of friction. To do this, we consider: (1) decision costs imposed by formal voting procedures: actors or institutions with veto power at any point in the parliamentary process; (2) other decision costs such as those associated with the capacity of some institutions, like

Table 2.3 Institutional friction associated with different policy-making channels

	Policy-making channel	Initiator	Formal voting required	Restrictions on issue content	Other decision costs	Institutional friction
Speeches	State of the nation	Prime Minister	No. Only the proposals for resolution introduced by parliamentary groups following the debate are voted.	No formal restrictions	No	Very low
Speeches	Investiture	Candidate to Prime Minister	Yes. The political program presented by the candidate requires the support of the Congress in a final voting (overall majority in a first voting and minority of the members in a second voting)	Informal restrictions apply in relation to the content, especially when there is not absolute majority in the Chamber.	No	Low
Oral questions	Oral questions in plenary meetings	Members of Parliament (MPs)	No	Formal restrictions apply: oral questions must be relevant, fall directly under the responsibilities of the Executive, and never deal with personal or private matters or disorderly expressions.	The Parliamentary Bureau qualifies the questions: quota by parliamentary group and issue content. The Board of Spokespersons, with the agreement of the government, can decide on the substitution of questions already in the agenda or on the reduction of the total number discussed per session.	Low

Continued

Table 2.3 Continued

	Policy-making channel	Initiator	Formal voting required	Restrictions on issue content	Other decision costs	Institutional friction
Bills	Executive bills	Executive	No to enter the parliamentary procedure. Yes to be passed (vary according to the legislative procedure).	No restrictions	The Parliamentary Bureau orders their publication in the Official Journal of the Chamber and sets the deadline for tabling amendments.	Moderate
	Parliamentary bills	14 MPs; 25 senators; parliamentary groups with the signature of the group's speaker; regional parliaments'Popular initiative	Yes to enter the parliamentary procedure: taking-into-consideration debate and formal voting in the Chamber (with the exception of those coming from the Senate). Yes to be passed (vary according to the legislative procedure).	Cannot be related to budget issues nor involve credit increase or income reduction. Those coming from Popular initiative cannot be related to issues reserved to organic law, taxes, international affairs nor to prerogative of mercy.	Those bills coming from the Congress (MPs) once published in the Official Journal of the Chamber are sent to the government that has to agree with the initiative. In the case of Popular initiative, qualification by the Parliamentary Bureau.	Moderate

	Decree-laws	Executive	Yes, they have to be submitted for debate and voting by the entire Congress within 30 days of their promulgation. Congress has to adopt a specific decision on their ratification or revocation, having the possibility to process them as executive bills.	Can only deal with issues not related to the regulation of basic institutions, the CCAA, and fundamental rights or freedoms regulated in Title I of the Constitution.	No	Moderate
	Legislative-decree	Executive	Require a previous delegation act (with law status) issued by the Congress.	Issues not reserved to organic law.	No	Moderate
Laws	Ordinary laws	Bills	Yes	Issues not reserved to organic law.	No	High
	Organic laws	Bills	Yes, including a final voting over the total text that requires absolute majority of the Chamber.	Limited to the regulation of fundamental rights and public liberties; the national electoral system; approval of regional statutes; basic institutions; states of alert, emergency and siege; and maximum structural deficit.	The Parliamentary Bureau and the Board of Speakers qualify the law as organic.	Very high

Source: Reglamento del Congreso (www.congreso.es) and *Boletín Oficial de las Cortes Generales* (BOCG).

the Parliamentary Bureau or the Board of Speakers, to qualify initiatives guaranteeing that they fulfill the formal requirements established by the rules of the Chamber; and (3) issue content restrictions: to what extent some of these agendas are limited by formal rules regarding the regulation of some topics to the exclusion of others.

This ranking provides a starting point for analyzing how agenda capacity varies across policy-making channels. Agenda capacity has been measured simply considering the total number of laws, bills, decree-laws, legislative-decrees and oral questions introduced in the Spanish Parliament (and the mean by legislature), the total number of media stories in the front pages of *El País* and *El Mundo*, and the total number of quasi-sentences in the Prime Minister speeches and party manifestos. As Table 2.4 illustrates, those agendas associated with lower institutional friction have higher agenda capacity (Baumgartner and Jones, 2002; Talbert and Potoski, 2002. The capacity of the Spanish Parliament to process organic laws is lower than in the case of ordinary laws. If we consider the total legislation passed from 1980 to 2011, 15 percent correspond to organic laws and 61 percent to ordinary laws.

This is mainly explained because organic laws require the agreement of the absolute majority of members in the Chamber to be passed and they are also subjected to limitations in terms of issue content. Organic laws can be used only for the regulation of fundamental rights, the general electoral system, the approval of regional statutes (Estatutos de Autonomía), basic institutions like the Constitutional Court or the Ombudsman (Defensor del Pueblo), and the states of alert, emergency or siege. Since the Constitutional reform introduced in 2011, an organic law regulates also the maximum structural deficit that the state and regional authorities may have in relation to their gross domestic product. The Parliamentary Bureau[10] and the Board of Spokespersons[11] are responsible for certifying laws as organic.

The capacity of the executive to process decree-laws is also low compared to executive bills (they represent 21 percent of the total legislation). Decree-laws are provisional regulatory acts passed by the executive in case of extraordinary and urgent need, when exceptional circumstances make following ordinary legislative procedure impossible. They are also subjected to issue content restrictions as they can only deal with issues unrelated to the regulation of basic institutions, the regions, or fundamental rights or freedoms regulated in Title I of the Constitution. They are provisional because they have to be submitted for debate and voting by the entire Chamber of Deputies within 30 days of their promulgation. The Chamber has to adopt a specific decision

Table 2.4 Summary of the Spanish Policy Agenda databases

Type of agenda	Capacity	Diversity	Volatility
Government			
Investiture speeches	2,521[a]	2.5	8.2
Annual speeches	7,528[a]	2.5	7.4
Executive bills	1,305	2.6	5.1
Decree-laws	384	2.6	6.8
Parliament			
Organic laws	274	1.8	10.6
Ordinary laws	1,123	2.6	5.1
Parliamentary bills	2,008	2.6	5.7
Legislative-decrees	54	2.4	12.6
Oral questions plenary meetings	15,238	2.8	3.4
Party manifestos			
PSOE	16,688[a]	2.8	3.1
PP	18,570[a]	2.9	2.9
Media front pages			
El País	56,540[b]	2.7	4.7
El Mundo	50,766[b]	2.5	5.7

Note: Period covered is 1982 through 2011, with the exception of media front pages (1996–2011) and public opinion (1993–2011);[a] Number of quasi-sentences;[b] Number of stories in the front pages.

Source: Spanish Policy Agendas Databases (www.ub.edu/spanishpolicyagendas), *Congreso* databases (www.congreso.es), and *Centro de Investigaciones Sociológicas* (www.cis.es).

on their ratification or revocation in the same period, having the possibility to process them as executive bills. The number of legislative-decrees passed is even lower (3 percent of the total legislation). Legislative decrees are elaborated by the executive after a delegation act is issued by the Parliament and, like decree-laws, can only be related to issues not reserved to organic law. The Parliament must define precisely the purpose and scope of the legislative delegation, as well as the principles and criteria to be followed in its application.[12]

Decision costs explain why, even if parliamentary groups introduce more bills than the executive, parliamentary bills have almost no chance to go through the legislative process. The formal rules that govern the pass of legislation in the Chamber give executive bills priority in front of other types of bills so that most Spanish laws have their origin in governmental initiatives. As Table 2.3 illustrates, a bill introduction is a quite open process in which several policy actors can participate – the executive, parliamentary groups of any of the chambers, the general population, and regional authorities. They can be introduced by 14 deputies, 25 Senators, by a parliamentary group with the signature of the speaker, by regional parliaments or by popular initiative following a quite restrictive procedure that requires 500,000 signatures of support. Parliamentary bills have to pass a taking-into-consideration debate and a formal voting in the Chamber (with the exception of those coming from the Senate) so that only those having a wide support have the possibility to enter the legislative process.

These institutional constraints do not exist for executive bills; the only requirement in that case is that they be passed in the Council of Ministers.[13] Executive bills are initiated within a ministerial department, and once passed in the Council of Ministries they go directly to the Parliament where after publication in the *Boletín Oficial de la Cortes* they initiate the formal legislative procedure. These variable decision costs explain why 90 percent of the bills introduced by parliamentary groups fail, but 92 percent of the executive bills are eventually enacted into law. Parliamentary bills are subject also to issue content restriction, as they cannot be related to the budget, increase spending, or decrease revenues for the state. Those that originate in popular initiatives cannot be related to issues reserved for organic laws, that is, the organization of state functions, taxes, international affairs, or criminal clemency procedures. The capacity of the government to declare executive bills urgent, reducing the length of the legislative procedure, explains why they are processed more rapidly than bills introduced by parliamentary groups (respectively, 234 days versus 388, on average).

Introducing an oral question entails decision costs, but these are far lower than those associated with passing a bill. Oral questions introduced in plenary meetings are not subjected to any formal vote, but still some institutional constraints exist. Restrictions on issue content apply to all parliamentary groups. Oral questions must be relevant, fall directly under the responsibilities of the executive, and never deal with personal or private matters, or use disorderly expressions. The existence of a quota of questions attributed to the different parliamentary groups for each session according to their number of seats significantly reduces the capacity of small parties to introduce oral questions. The capacity of the Board of Spokespersons (with the agreement of the government) to decide on the substitution of questions or to reduce the total number of questions discussed per session introduces also a decision cost that does not exist for other agendas. As a result, the overall number of questions introduced in plenary meetings by legislature is very high. Chapter 4 shows that there is wide variation in the total number of questions introduced over time and across parliamentary groups, depending on the formal rules that regulate the introduction of questions, whether groups are in government or in opposition, and party preferences.

Speeches and manifestos are the policy-making channels associated with lower levels of friction and higher capacity. In the case of speeches, there is not a defined limitation in their length and the costs of shifting attention to new issues are relatively low compared to other types of activity. If the Prime Minister wants to give more attention to an issue, he does not need the consent of other political actors as is required, for example, to pass a law. In the case of the investiture speech, there is a final vote in the Chamber of Deputies that introduces an institutional cost that does not exist in the case of the annual speech. As a result, informal restrictions apply in relation to the content, especially when there is not absolute majority in the Chamber. In the case of the *investiture speech*, the Chamber rules explicitly say that the speech of the candidate has no time restrictions. The annual debates are not formally regulated in the Constitution nor in the Chamber's rules – the debate was established by Prime Minister Felipe González in 1983 and follows the Chamber's rules provisions regarding governmental communications (*comunicaciones del gobierno*). The debate usually takes place over two days. The Prime Minister initiates it with a general intervention in the morning session, without time restriction. Following the interventions of the different parliamentary groups, these can present proposals for a resolution, which are debated and voted in a final session.

Party manifestos have low friction and high agenda capacity because political parties can make promises and write statements on a wide range of issues without any time or space restrictions. Compared to other policy-making channels, however, they have a very low visibility. Although the definition of a party's manifestos is still among the most important tasks to be performed by political parties when election time comes, the number of people who actually read them is very small. On the other hand, speeches are one of the most visible political moments of the year, mainly because what the Prime Minister says is reported by the media and watched and debated by the public and their representatives.

Finally, the capacity of the media to process information is limited because the range of issues that might be covered in the news is always much greater than the available space front pages allow (Boydstun, 2013). *El País* and *El Mundo* publish an average of ten and nine stories per day in their front pages, respectively. As a result, newspaper editors have to go into an important process of issues prioritization, picking and choosing among the thousands of stories that are generated daily, from events in society to day-to-day politics. Despite this, as we illustrate in the next section, the agenda of these media outlets is strongly concentrated on only some topics (Chaqués-Bonafont and Baumgartner, 2013).

Agenda diversity

Agenda capacity has an inherent upper limit, but with the amount of time and resources available, policy actors can focus their attention on just a few affairs or distribute it more equally across a large set of issues. The concept of *agenda diversity* describes how attention is distributed across issues providing a measure of agenda variation (McCombs and Zhu, 1995; Jochen and De Vresse, 2003; Jennings et al., 2011). More issues mean that it is more diverse and complex, and fewer issues mean that it is more homogeneous and uniform. To measure agenda diversity, we have calculated the Shannon entropy index (see Table 2.4). This index measures the distribution of attention across the 19 CAP issue categories, providing an indicator of the relative concentration or dispersion of data (Baumgartner et al., 2000; Talbert and Potoski, 2002; Wolfe, 2008; Boydstun et al., 2014). It ranges from 0 to the natural log of 19 (3). A score of 0 indicates that attention is concentrated in a single topic; by the contrary, a score of 3 indicates that attention is perfectly distributed across issues. The higher the entropy, the less concentrated the agenda.[14]

The Shannon index can be considered the opposite of Concentration. If all the attention were focused on just one topic, to the exclusion of all others, we would say that attention was highly concentrated. If attention were equally spread across all possible topics, concentration would be at its lowest possible value. The index, then, is the degree of spread in attention. Low scores mean highly concentrated attention. High scores mean great diversity or spread in attention.Table 2.4 illustrates that agenda diversity is strongly related to agenda capacity and institutional friction (Baumgartner and Jones, 2002; Talbert and Potoski, 2002; Green-Pedersen, 2007). Those policy-making channels that have more capacity to process information, like oral questions and party manifestos, are more diverse than those associated with higher institutional constraints. Attention is especially concentrated in organic laws, where only two issues – justice and governmental affairs – represent more than 50 percent of the total agenda (see Table 2.5). By its very nature, this policy-making channel is also among the most used for the regulation of rights and defense. Other substantive agendas associated with more moderate friction – ordinary law and executive bills – are much more fragmented than organic laws but still highly concentrated

Table 2.5 Percentage of attention across topics and policy-making channels

Topic	Organic law	Ordinary law	Decree-law	Legislative decree	Parliamentary bills	Executive bills	Oral questions plenary	Party manifestos	Annual speech	Investiture speech	MIP	*El País*	*El Mundo*
Economy	4.6	19	17.3	16.7	6.3	17.1	6.9	7.6	18.6	16	40.4	5	4.1
Rights	9.2	1.7	.5	0	8.7	3.5	3.3	3.8	3.4	4.6	6.8	8.5	7.6
Health	.8	2.7	3.8	0	1.9	2.3	6	4.8	1.5	1.8	7.4	3.9	3.2
Agriculture	0	3.8	6.6	3.7	1.7	3.5	7.1	3.3	1.1	.4	.5	1	.8
Labor	1.1	5.6	10.4	11.1	7.9	4.5	4.4	7.8	12	7.1	1.9	2.4	2.2
Education	4.2	4	3.3	1.9	6.2	4.8	8.1	12.4	4.8	7.4	1.5	3.7	3.6
Environment	0	2.4	4.4	5.6	2.4	2	2.7	2.6	.8	1.3	1	1.6	1.1
Energy	0	2.2	3.3	1.9	.6	2	3	1.8	.7	.6		2.2	1.9
Transportation	1.5	6.3	6.6	11.1	3.1	5.9	9.1	4.5	1.7	2	.2	3.5	2.7
Crime & justice	29.9	8.7	4.4	3.7	14.5	12.8	9.5	7.4	12.5	10.8	23.1	19.9	25.2
Social policy	0	2.1	.8	0	7.3	.9	2.2	6.4	2.3	3.5	3.4	1	.7
Housing	0	1.4	.5	5.6	2.5	1.3	2	2.5	1.6	2	4	1.2	.7
Commerce & banking	.8	13.7	18.1	18.5	6.3	12.5	4.5	6.1	2.7	1.3		4.7	4.1
Defense	7.3	3.6	2.5	1.9	5	4.5	5.1	2.3	4.4	3	.4	8.1	7.2
Science & technology	0	3.9	2.7	1.9	3.6	3.5	3.9	5.1	1.9	2.8		3.6	3
Foreign trade	.8	.9	1.1	1.9	.4	.8	.9	1.6	2.4	1.3		1.1	.8
International affairs	6.1	6	1.1	0	1.2	6.5	6.7	7.5	11.5	14.2	0	7.1	5.4
Government	33.7	8.4	5.8	13	18.4	8.9	11.8	11	15.5	18.9	9.2	21.1	25.5
Public lands	0	3.3	6.6	1.9	2.1	2.7	2.7	1.4	.8	1	.2	.4	.2
Total	100	100	100	100	100	100	100	100	100	100	100	100	100
Topics representing 50% of the total	2	4	4	4	4	4	6	6	5	4	2	3	3

Note: Data cover the period of 1983 to 2001, with the exception of front page newspaper stories (1996–2011) and public opinion (1993–2011). The percentages sum to 100 percent down the columns, therefore indicating the relative share of attention across the topics. The last row shows the number of topics that comprise half of the total. This is a good measure of concentration, as law numbers indicate that a majority of attention is focused on just a few topics.

Source: Spanish Policy Agendas Databases (www.ub.edu/spanishpolicyagendas).

on the regulation of basic functions such as the economy and industrial policy, government operations, law and order, defense, or international relations. The consolidation of the modern welfare state has led to expansion of the traditional functions of states to new issues like health, education, or social policy, but the governmental agenda is still very concentrated on these core functions (Jennings et al., 2009).

Decree-laws are used for the regulation of a broad range of issues but especially those related to the economy, industrial policy, and labor (Table 2.5). Although they should be used only under exceptional circumstances, decree-laws have become an important regulatory instrument in the hands of the executive to bypass time-consuming parliamentary procedures and also to avoid the blocking of proposals by opposition groups in controversial issues, especially when the incumbent does not have the absolute majority of seats in the Chamber. They are especially used by the government in times of economic crisis to pass, for example, labor market reforms, policies to promote job creation, or to redesign unemployment benefits. As we explain in Chapter 3, the use of decree-laws has increased significantly over time, and they are used to make decisions that have nothing to do with urgent necessities in a context of increasing polarization and lack of consensus in Spanish politics.

Speeches have more capacity and lower decision costs than substantive agendas, which might be expected to bring about a more diverse distribution of attention across issues, but despite this they are very concentrated also on core functions. The Prime Minister mainly focuses on the economy, government operations, justice and law, and international affairs (these four topics account for more than 50 percent of the total investiture and annual speeches). No important differences exist between the two types of speeches, with the exception of labor issues that capture a higher percentage of attention in the annual speech, making the agenda slightly more diverse. Speeches are highly concentrated because they focus on issues with a preeminent status in politics (Jennings et al., 2009) and because, as one of the most visible moments of the year, the Prime Minister uses them to respond to the main priorities of public opinion, especially variations in the economic situation (see Chapter 9).

As Table 2.5 illustrates, the agenda of the public mainly focuses on the economy (40 percent of the total), an issue that has been traditionally the most important problem for Spanish citizens. The economy has always represented an important part of the Prime Minister speeches, especially during periods of recession and high unemployment. That was the case in the mid-1980s when, following transition to democracy and after a long period of growth, the Spanish economy entered a crisis, leading unemployment to reach 22 percent of the total active population. In the early 1990s, there was a new economic phase of deterioration and growing unemployment, but after the mid-1990s, Spain enjoyed rapid economic growth – even leading to government surplus from 2005 to 2007. As we illustrate in Chapters 3 and 9, the government of Rodríguez Zapatero that was governing after 2004 paid the least attention to the economy of all previous governments. But, following the outbreak of the current economic and financial crisis, economic issues have emerged to capture the highest share of attention of the democratic period. Issues, such as regulations of the real-estate market, banking policy, the international financial system, and bankruptcy, that had never or rarely been present in a Prime Ministerial speech, became top agenda items.

Internal affairs and justice, mainly terrorism, have been the second-most important problem for Spanish citizens across the period of our study, with 23 percent of MIP responses, and account for a large share of the Prime Minister's attention in both the annual and the investiture speech. Spain has seen many brutal and deadly terrorist attacks committed by the Basque organization ETA, responsible in Spain for over 3000 violent acts, including over 50 kidnappings, almost 800 murders, and many more cases of extortion, blackmail, and intimidation (Gunther and Montero, 2009); but also by Islamist groups linked to Al Qaeda, responsible for the massive terrorist attack that occurred in Madrid just three days before the 2004 election, which killed nearly 200 people. These terrorist actions bring about great media, public, and political attention, having important concentration effects on the agenda.

Governmental affairs, and more specifically issues related to political decentralization and corruption, also represent an important part of the public and the Prime Minister agenda. The debate surrounding the *Estado de las autonomías* captured an important part of attention during the early 1980s – in 1983, for example, almost 20 percent of the speech focused on this topic – and during the 1990s, as a result of the debate over the homogenization of jurisdiction among regional authorities. More recently, the decision of some regions to reform their statutes of autonomy and the independence movement in Catalonia have reopened the debate over the territorial structure of the state, capturing an important share of the public and political agenda. In the case of corruption, this issue concentrated the Prime Minister speech especially in the last legislature of González, as a result of different political and economic scandals that affected the PSOE in the early 1990s, and that seriously affected government's credibility. In 1995, corruption represented 33 percent of the total speech and was the second-most important problem for Spanish citizens, even above terrorism.

Issues rarely mentioned by citizens as the most important in Spain, like those related to science and technology or energy, are not mentioned by the Prime Minister, either, with the exception of international affairs. Speeches are used to respond to public concerns but also to define and explain the executive's position on issues that are high on the agenda but not formally regulated by law, like international relations. Prime Ministers have used speeches to justify decisions on defense and international relations, from EU politics to involvement in military conflicts. For example, Aznar devoted an important part of its annual speech in 2003 to legitimate military involvement in the Iraq war. Zapatero also gave high priority to this issue in its 2004 investiture speech, but in this case, to justify the removal of troops.

The low level of institutional friction associated with speeches allows the Prime Minister to react to signals coming from the environment, but as they are given only occasionally – at the beginning of the term in the case of investiture speeches and once a year in the case of the state-of-the-nation debate – the government uses other policy-making channels that also have high levels of visibility, like oral questions introduced in plenary meetings, to react to unexpected events or to attract the attention of the public. Opposition groups use also oral questions to bring some issues to the consideration of the public, especially those that are important to their voters, and to give visibility to policy failures or government mismanagement. Because they serve different purposes and goals and they are subjected to low levels of friction, oral questions are among the most diverse policy-making channels. They focus on core functions like the economy, governmental affairs, and crime and justice, but also leave space to more selective issues like transport, energy, or agriculture. As Table 2.5 illustrates, in the case of oral questions, we have to consider up to six topics to account for 50 percent of the total agenda.

Parliamentary bills are formally designed to pass legislation, but as a result of the high decision costs imposed by the parliamentary procedure, they are also used by opposition parliamentary groups as attention-getting devices, performing a similar role to non-legislative acts. Like other policy-making channels, they focus on basic functions (e.g., internal affairs and justice and government operations), but they are less focused on core issues (e.g., the economy) and more focused on rights and social policy. Although formally designed to pass legislation, parliamentary bills have strong symbolic content and are used by opposition groups to draw attention on the need for new legislation on issues like abortion, housing, or immigration.

The high agenda capacity and low level of friction of party manifestos turns them into one of the most fragmented policy-making channels. Basic functions, like those related to the economy, governmental operations or international affairs, represent an important part of the symbolic agenda of political parties; but compared to other policy-making channels, they pay much more attention to welfare issues, and especially to education and social policy. If we compare the manifestos of the two main parties, the PSOE and the PP, results illustrate that the conservatives pay slightly more attention to issues like the economy, health, agriculture, energy, transport, and justice and internal affairs or defence, and the socialists to rights, labor, social policy, education, housing or international affairs, but differences are low and not statistically significant in most cases.

Finally the media agenda in both *El País* and *El Mundo* is also very concentrated on few topics. As Chaqués-Bonafont and Baumgartner (2013) illustrate, although the Spanish media system is highly polarized, both newspapers pay attention to the same issues, mainly justice and crime (terrorism), government operations (political parties, campaigns, and corruption) and rights (immigration and rights related to voting and political participation).[15] As Table 2.4 illustrates, these three topics account for more than the 50 percent of the total media agenda. Although there are some differences between the papers – more internal affairs in *El Mundo*, for example, or more international affairs in *El País* – readers of the two papers are getting generally a very similar mix of stories (Chaqués-Bonafont and Baumgartner, 2013: 73).

Agenda volatility

Beyond the capacity of each agenda, and related to the issues of diversity discussed above, institutional agendas also differ in their volatility. Several recent studies have argued that the press of more issues to capture policy makers' attention has resulted in a more volatile and competitive agenda (McCombs and Shaw, 1972; McCombs and Zhu, 1995; Talbert and Potoski, 2002; Jones and Baumgartner, 2005; Green-Pedersen, 2007). Agenda volatility is conceptualized as the "life span" of an issue, or the rate at which issues are replaced after emerging on the agenda. It measures whether they stay for long periods of time or, on the contrary, they come and go rapidly. Sensitivity to a greater range of issues does not by itself produce volatility, but when a greater number of issues are competing for attention on an agenda that has major constraints on its capacity, volatility is a likely outcome (McCombs and Zhu, 1995: 497, Jones and Baumgartner 2005).

Volatility has been measured in the following way. For each issue we have calculated the percentage of attention by topic and legislature (to avoid the deviations introduced by election years when the parliamentary activity decreases) and calculated the standard deviation. In the case of public opinion, speeches, and the media, we have considered the yearly percentage of attention by topic. Results illustrate that in agendas with high capacity and low institutional friction, like party manifestos and oral questions, volatility rates are very low (see Table 2.4). These agendas are very fragmented and show similar patterns of attention distribution to different issues over time. Speeches and the media have a more concentrated agenda, combined with moderate levels of volatility. In the case of executive bills and ordinary laws, volatility is moderate, as they concentrate on the regulation of basic functions and day-to-day policies with little variation over time. By contrast, agendas associated with more institutional constraints, like organic laws, have higher volatility. In the case of decree-laws and legislative-decrees, issues also enter and leave the agenda rapidly because they are mainly used by the government in times of economic crisis or in other periods when government feels the need to respond quickly to changing circumstances or unexpected events, even though decrees are increasingly used as ordinary policy-making instruments (see Chapter 3).

Conclusions

The Spanish Policy Agendas Project databases consist of 11 comprehensive, reliable, and comparable data sets that record the issue priorities of the government, parliament, the political parties, the public, and the media over a long time period. All the series have been coded according to the coding scheme developed by the Comparative Agendas Project so that, for the first time, we have reliable data on many aspects of the policy-making process that can be systematically compared across data sets with confidence. This high-quality database can be used to study aspects of the Spanish political system in a manner hitherto impossible. For example, we can develop detailed studies of particular policy areas, analyze the capacity of different agendas to respond to signals coming from the environment, the correspondence between the policy priorities of the public and the promises of party manifestos, to what extent the policy statements of the Prime Minister speech translate into policy outputs, or the capacity of the media to influence the public or the political agenda. The development of a specific set of indicators designed to measure the delegation of political power upwards toward the EU and downwards toward regional authorities allow analyzing, from a completely new perspective, how the consolidation of a multi-level system of governance has affected agenda dynamics in Spain.

In this chapter we have explained the content and logic of these databases, but also provided a first general description of the institutional friction associated with the different parts of the policy process and analyzed its implications in terms of agenda capacity, diversity, and volatility. Results illustrate that those agendas with higher friction have less capacity to process information than those associated with lower costs. Overall, substantive agendas designed to pass formal legislation have lower agenda capacity than those with more symbolic content, like speeches, oral questions, and party manifestos. The agenda is also much more diverse in those policy-making channels associated with low institutional friction, with the exception of speeches. Substantive agendas concentrate on the regulation of basic functions, like the economy, government operations, crime and justice, international affairs, and defense. Speeches, contrary to party manifestos or oral questions, with higher capacity to distribute attention across a broad range of issues are not very fragmented, but also focus on core issues and especially those that are closer to the priorities of public opinion. Similarly, the media agenda is also concentrated in very few topics. Results show also that those agendas with lower capacity are generally also more volatile, with agenda items entering and leaving the agenda more rapidly.

In the various chapters of this book, we use this set of databases to analyze how policy issues get into the agenda, considering as explanatory variables institutional factors, focusing events, and new flows of information. We start by considering to what extent partisan politics explain the pattern of prioritization of issues by different governments in Spain from transition to democracy to present, and by testing to what extent there is a connection between the priorities of the executive and the promises defined in the party manifestos, which led to their election in the previous campaign.

Part II

The Dynamics of Political Institutions in Spain

3
Broken Promises

Government partisanship is generally understood as one of the key variables explaining policy change (Budge and Keman, 1990; Petrocik, 1996; Walgrave and Nuytemans 2009). Political parties prioritize different issues and policy proposals according to their preferences, and once they win elections they try to implement them, following their mandate. Thus, social democratic and conservative parties would adopt different strategies to respond to economic crises (Boix, 1998; Manin et al., 1999) or moral issues (Engeli et al., 2012; Varone et al., 2008; Chaqués-Bonafont and Palau, 2012). The left would prioritize increases in tax rates and public expenditures as a means to reconcile growth and equality, while conservatives would attempt to keep taxation low, promoting free-market strategies. Left parties would push forward regulatory change on moral issues (i.e., abortion, stem-cell research, and same-sex marriage), toward increasing liberalization, while conservative parties would try to promote less permissive policies. Differences would also be expected between statewide and regional parties in relation to regionalization. Basque and Catalan parties would always push for increasing political autonomy to the regions, while the PP and PSOE would defend the power of the central state, whether in maintaining the status quo when others push to decentralize, or by promoting recentralization if their rivals have previously had success.

Party preferences explain how governmental actors prioritize issues in Spain. However, preferences alone cannot explain why Felipe González (PSOE) did not push forward the regulatory reform of abortion in the early 1990s, why regional fiscal reform was one of the very first decisions taken by the minority government of José María Aznar (PP) in 1996, why the economy displaced social and civil rights on the agenda of Rodriguez Zapatero, or why the policy responses of the PP and PSOE to the economic recession of the late 2000s were almost identical. The capacities of parties to implement their policy programs and to respond to their mandate while governing depends to a large extent on institutional factors, focusing events, and flows of information often reflecting unanticipated events outside of the control of the government (Baumgartner and Jones, 2015).

Institutional factors help explain how governmental actors define their goals and strategies over time. Formal rules account for the predominant position of the Spanish executive in front of the legislature at the national and EU level (Döring, 1995; Moravsick, 1994; López-Nieto, 2001; Capo, 1994). They also guarantee the formation of stable, single-party governments that, for more than half of the democratic period after transition to democracy, have ruled with the majority of seats in Parliament. In this context, governments' institutional capacity to implement their mandate is very high, with some variations across time, depending on institutional factors like the type of government (that is, the government's majority/minority status) and also unexpected events like an economic crisis.

Our results indicate that under minority status and/or bad economic conditions the executive tends to respond less to its electoral promises and has fewer chances to pass executive bills into laws. In this regard, we find support for both mandate theory (Budge and Hofferbert, 1990; Klingemann et al., 1994) and the policy dynamics approach (Baumgartner and Jones, 1993; Jones and Baumgartner, 2005). Unexpected events like the fall of the Berlin Wall in 1989, German reunification, or the 1991 Iraq war are key factors explaining part of the governmental agenda of Felipe González (Barbé, 1999). Environmental catastrophes like the Prestige oil spill in 2002 or food-safety outbreaks like the mad-cow scandal of 1996 and beyond explain why issues like food safety had a predominant position in the agenda of José María Aznar (Palau, 2009). By the same token, the economic recession of the late 2000s led to a sudden increase of economic issues into the agenda of José Luis Rodríguez Zapatero and the displacement of social and civil rights off the policy agenda. These unexpected events were impossible to predict, and thus they were never mentioned as key issues along the electoral campaign. Still, once they happened, policymakers had to respond to them, giving these issues a disproportionate share of the governmental agenda even though they were not the stated priorities of the winning party in the previous elections.

However, we demonstrate in this chapter that the type of government or bad economic conditions alone cannot explain why the governing party is increasingly governing by decree, ignoring the role of the

Spanish Parliament in the decision-making process, nor why mandate responsiveness is declining across time. For the last decade, the number of decree-laws has increased exponentially. Organic laws and decree-laws have been increasingly passed despite the negative votes of opposition parties. Further, most ordinary laws have not been debated in plenary meetings in Parliament because of the increasing use of non-ordinary procedures to pass bills into law. These changes in the policy style are not related to a transformation in formal rules (decree-laws are regulated by the Spanish Constitution in 1978) or economic conditions (there were severe economic crises in the early 1980s, in the 1990s, as well as more recently). Rather, they relate to changing party preferences toward a new way of doing politics characterized by increasing confrontation and lack of consensus.

Finally, we argue the gradual decline in government's capacity to put into practice its stated policy preferences is linked not only to the type of government and economic conditions but also to the consolidation of a multilevel system of governance (Chaqués-Bonafont et al., 2014). The gradual delegation of authority upwards to the EU and downwards to the regions limits the capacity of national governments to define their agendas according to their preferences. Governmental actors lose some of their predominant position in the policymaking process as issue jurisdiction becomes increasingly fragmented, shared or divided across different levels of government – supranational, national, and subnational. In this multilevel system of governance, the party in government may find important barriers – these may be called transaction costs or institutional friction – to implement electoral promises about issues whose jurisdiction is shared with the regions and/or the EU. By the same token, governing elites may have to push forward bills transposing EU directives about issues that were not taken into account during the electoral campaign simply because they were not directly affecting Spanish national interest, like regulations to stop the commercialization of products derived from endangered harp and hooded seal pups.[1] In this regard, at the end of this chapter we provide evidence of how mandate responsiveness declines especially for those issues whose jurisdiction is shared with the regions and with the EU.

To explore these issues, the first part of the chapter explains variations in the content of the governmental agenda across time, political parties in control of the government, and individual Prime Ministers. We conclude that issue attention varies significantly between political leaders, but these differences are not so apparent at the party level. From here, we focus on variations in the use of legislative instruments by the executive, emphasizing how increasingly it takes decisions which exclude the Parliament from participation, contributing to confrontation and polarization of the political debate in the last decade. Finally, we explain how multilevel governance and shared jurisdiction have played an important role in the decline of mandate responsiveness across time.

Changes in the governmental agenda

Issue attention in the Spanish executive branch varies across time in both the symbolic and decisional agenda. While giving their speeches or drafting legislative proposals, governmental actors pay increasing attention to commerce and banking, rights, the environment, crime and justice, education, and social policy (see Table 3.1). By contrast,

Table 3.1 Time effect on attention to different topics

	Executive bills			Speeches		
Topic	**B**	**Sig.**	**R^2**	**B**	**Sig.**	**R^2**
Economy	.048	.808	.002	.062	.687	.006
Rights	.098	.359	.047	.239	.043	.190
Health	.016	.826	.000	-.023	.618	.013
Agriculture	.081	.100	.124	-.026	.280	.068
Labor	.146	.123	.115	-.083	.699	.006
Education	–.167	.163	.090	.146	.047	.155
Environment	.194	.036	.279	.021	.737	.010
Energy	–.049	.441	.050	.032	.109	.162
Transportation	–.167	.125	.099	.048	.352	.043
Crime & justice	.279	.013	.222	–.008	.960	.000
Social policy	.027	.541	.048	.105	.031	.186
Housing	–.073	.598	.026	.109	.215	.094
Commerce	.512	.002	.306	.208	.061	.181
Defense	–.184	.037	.200	–.269	.000	.511
Science & technology	.022	.766	.005	.060	.201	.085
Foreign trade	.001	.987	.000	-.037	.314	.040
International affairs	–.212	.053	.160	–.516	.003	.289
Government	–.165	.223	.061	–.483	.028	.166
Public lands	–.011	.877	.002	.052	.036	.296

Note: The dependent variable is the percentage of attention to the topic listed. The independent variable is time (years, from 1983 to 2011 in the case of executive bills and from 1982 to 2011 in the case of speeches). Positive B-coefficients indicate topics trending toward more attention over time, and negative coefficients show decreasing attention over time. The level of significance show whether these trends are statistically significant or not (Sig. should be below .05 to indicate a statistically significant trend). The models were run with a constant term, but these are not shown here.

attention to defense, international affairs, and governmental issues decline gradually from 1982 to present, while for some other issues like health, agriculture, labor, and macroeconomics attention follows a more random pattern. These changes in issue attention are not only explained by partisan logics, but the preferences of the head of the executive and reactions to events outside of their control, mainly the economic crisis.

Tables 3.2 and 3.3 show mean differences in the pattern of issue prioritization across the governments of González (1982–1996), Aznar (1996–2004), and Rodríguez Zapatero (2004–2011). While giving his speeches, Felipe González pays significantly more attention to governmental issues (regionalization, foreign trade, entry into the EEC),

Table 3.2 Differences in legislative activity across topics, by prime minister

	González-Aznar		Aznar-Zapatero		González Zapatero	
Topic	Mean difference	Sig. (t-test)	Mean difference	Sig. (t-test)	Mean difference	Sig. (t-test)
Economy	−3.5	.324	5.7	.268	2.2	.565
Rights	−.2	.874	−2.5	.367	−2.7	.271
Health	−2.3	.073	1.6	.314	−.7	.439
Agriculture	−1.6	.123	.8	.578	−.8	.358
Labor	−0.3	.870	−2.4	.323	−2.7	.168
Education	4.0	.101	.8	.668	4.7	.075
Environment	−.8	.297	−3.0	.131	−3.7	.064
Energy	1.3	.528	−.6	.614	.7	.626
Transportation	0.7	.747	2.2	.387	3.0	.226
Crime & justice	−1.2	.442	−5.1	.081	−6.3	.012
Social policy	−1.0	.365	.4	.828	−.6	.369
Housing	1.0	.639	−.9	.522	.2	.963
Commerce & banking	.9	.686	−12.2	.003	−11.4	.001
Defense	2.4	.196	−.3	.818	2.1	.382
Science & technology	−.9	.470	.3	.865	−.6	.724
Foreign trade	.5	.480	−.7	.403	−.2	.878
International affairs	−.3	.871	4.9	.026	4.6	.077
Government	3.8	.103	−2.8	.376	1.0	.751
Public lands	.9	.500	.1	.967	1.0	.537
Total agenda	.4	.567	−1.6	.111	−1.2	.201

Note: Entries show the difference in percent of Prime Minister's executive bills allocated to each topic. So if González devoted 10 percent of bills to the economy and Aznar 13.5 percent, the difference is −3.5 percent. The associated t-tests show whether any differences are statistically significant (indicated by values less than .05).

Source: Spanish Policy Agendas Databases (www.ub.edu/spanishpolicyagendas)

Table 3.3 Differences in attention across topics, by prime ministers' annual speeches

Topic	González-Aznar Mean difference	González-Aznar Sig. (t-test)	Aznar-Zapatero Mean difference	Aznar-Zapatero Sig. (t-test)	González-Zapatero Mean difference	González-Zapatero Sig. (t-test)
Economy	1.8	.389	–.8	.853	.9	.829
Rights	–1.8	.158	–4.5	.116	–6.3	.028
Health	–2.3	.007	2.2	.010	–.1	.896
Agriculture	–.5	.288	1.3	.015	.8	.096
Labor	5.1	.290	–3.5	.182	1.5	.758
Education	–2.9	.039	–.8	.584	–3.7	.027
Environment	.6	.722	-.3	.716	.3	.837
Energy	–.4	.250	–.2	.711	–.6	.258
Transportation	.5	.600	-1.4	.051	–.8	.430
Crime & justice	–3.9	.245	3.7	.331	–.2	.955
Social policy	.1	.887	–2.1	.055	–1.9	.064
Housing	.6	.592	–2.3	.116	–1.7	.299
Commerce	–.9	.277	–2.8	.265	–3.7	.162
Defense	2.3	.133	2.2	.036	4.5	.003
Science & technology	–1.2	.298	–.2	.837	–1.4	.020
Foreign trade	1.9	.002	–1.0	.217	.9	.261
International affairs	3.1	.364	8.4	.002	11.5	.003
Government	6.6	.166	3.4	.380	10.1	.033
Public lands	–1.1	.023	–.1	.873	–1.2	.050
Total agenda	2.2	.017	.3	.702	2.5	.006

Note: Entries are similar to those in Table 3.2, but refer to the Prime Minister's annual speech.

Source: Spanish Policy Agendas Databases (www.ub.edu/spanishpolicyagendas).

and international relations than Rodríguez Zapatero and Aznar. In contrast, attention to rights (immigration, gender, and moral issues) social policy, and environmental issues are significantly more relevant in the agenda of Rodríguez Zapatero, and this is so especially in relation to the PSOE governments of Felipe González more than Aznar's. Finally, from 1996 to 2004, the PP governments of José María Aznar pay significantly more attention to health and agriculture than any other Prime Minister. In relation to legislative activities, differences are not as important as in the case of speeches. As Table 3.3 illustrates, with the exception of crime and justice (which was among the priorities of the governments of González and Aznar) and commerce and banking (which was mainly regulated by Rodríguez Zapatero), differences are not statistically significant.

These simple comparisons indicate that for many issues, differences in attention are not linked to party ideology but the preferences of the Prime Ministers, which adapt their political programs to changing social, political, and economic conditions and institutional factors. For example, the consolidation of a multilevel system of governance imposes opportunities and constraints to governmental actors. Democratization coincided with transformations in the scope and activities of the Spanish state, including marked increases in the size of welfare and regulatory infrastructure as well as a gradual delegation of governmental autonomy upwards to the EU and downwards to the regions. These transformations – in the functioning of political institutions, the scope of state intervention, and the actors involved in sharing and exercising power – have a direct impact on which issues get into the agenda and when they have the chance to be prioritized. For example, environmental issues got into the Spanish agenda with years of delay in relation to most EU countries and its regulation is mainly explained by demands coming from European integration (Börzel, 2000). Similarly, most of the regulation passed by Rodríguez Zapatero in his last year of government, including a constitutional reform to control public finance, cannot be explained without considering recommendations from EU institutions.

Changing religious values and the increasing secularization of society have also created new opportunities for those seeking to promote policy change, especially in relation to education, moral issues, or family policy (Brasslof, 1998; Requena, 2005; Montero, 1987; Chaqués-Bonafont and Palau, 2012; Bianculli and Jordana, 2013; Esping-Andersen, 2009). By the same token, in less than three decades Spain has been transformed from a country of emigrants to a country with an immigrant population of more than 10 percent, generating a radical transformation in the labor market as well as in social relations. Issues like the construction of mosques, labor conditions of immigrants, or the border fences in Ceuta and Melilla have become key issues in the governmental and parliamentary agendas as a result of this major social transformation.

In short, changes in the governmental agenda are linked to party preferences, and also institutional factors and major transformations in the social, political, and economic spheres. The following sections briefly describe these ups and downs of the governmental agenda. This is a starting point from which we further explain the transformation in the governmental policy style and changes in mandate responsiveness across time.

Macroeconomics, foreign trade, and banking

The economy is always a key issue in the agenda of any executive (Figure 3.1). However, each Prime Minister seems to be talking about quite different economic concerns. One of the main goals of Felipe González, right after winning the general elections in October 1982, was to fight against the economic crisis (more than 20 percent of Spaniards were unemployed), create a modern and fully developed fiscal system (completely underdeveloped under the Franco regime), and push forward the modernization of industrial production, fostering innovation and internationalization among Spanish firms. Most of these goals were directly linked with the entry of Spain into the EEC in 1986, which transformed the Spanish industrial model into a free-market economy. In the mid-1990s, the last government of Felipe González had to deal with a second economic crisis, with unemployment rates again above 20 percent. González put into practice a highly contested labor reform – which generated two general strikes in 1992 and 1994 – and started the process of privatization and modernization of the public sector.

Attention to the economy in the symbolic agenda declines from 1996 to 2008, that is, when economic conditions were extraordinarily positive, as Figure 3.1 displays. In contrast to Felipe González, Aznar's speeches focuses on economic successes, the expansion and increasing internationalization of economic activities, the benefits derived from the Economic and Monetary Union (EMU), and the importance of the process of privatization, deregulation, and liberalization of economic sectors, especially those related with telecommunications and energy. From the mid-1990s, Spain is growing faster than most advanced democracies, with unemployment rates for the first time close to 10 percent in the early 2000s. Economic growth was very much linked to the expansion of the construction and tourism sectors, and increasing internationalization of domestic firms, which gradually became multinational firms (Campa and Guillen, 1999; Guillén, 2005; Guillén and Garcia-Canal, 2010). The increasing demand from foreign investors, migrants, and the baby-boom generation, the availability of credit at low prices, and especially the policy decisions taken by the PP government of José María Aznar aimed at the deregulation of land-use planning, were key factors in the expansion of the construction sector and economic growth (Jordana, 2014; Fishman, 2012; Quaglia and Royo, 2014).

In 2007, Rodríguez Zapatero happily argued Spain was one of the wealthiest nations in the world, with the capacity to surpass Germany and Italy in terms of GDP per capita. Actually, in 2007, Spain was a

Figure 3.1 Attention to the economy

Note: Speeches: percentage of the annual Prime Minister's speech devoted to the economy. Mood: See description of this variable in Chapter 2; it combines public opinion polls on the state of the economy, the state of the political situation, how the government is managing public affairs, and evaluations of EU membership. Unemployment rate: actual rate. Data are yearly.

Source: Speeches: Spanish Policy Agendas Databases (www.ub.edu/spanishpolicyagendas). Unemployment rates: IMF (www.imf.org). Mood: Centro de Investigaciones Sociológicas (www.cis.es) and Eurobarometer (www.ec.europa.eu/public_opinion/index_En.htm).

country with governmental surplus (7.1 percent of the GDP), unemployment rates below 9 percent for the first time in the recent democratic period, and an economy growing faster than most other EU countries, at more than 3.5 percent. In this context, Rodríguez Zapatero enforced the restructuring of the banking system following most of the decisions taken at the EU level (Royo, 2013), the so-called new model of "sustainable economy" (later called *Ley-Sinde*), and fostered the redefinition of especially relevant economic sectors, from gambling, to the retail sector, and the SICAV (open-ended investment companies). The PSOE government was talking about the need to transform the economy into a new model based on innovation, respectful of the environment, and capable of guaranteeing full employment and welfare improvement for Spanish citizens. At the same time, the PSOE was reinforcing the SICAV, the paradigmatic example of the privileges of the wealthy (first regulated under the mandate of the PP in 2003).

Interestingly, Rodríguez Zapatero devoted less than 4 percent of his first speech in 2004 to talking about macroeconomics, and only 10 percent in his first speech of 2008, despite the fact that, at the end of his first mandate, it was evident Spain was getting into the deepest economic recession of the democratic period. After more than a decade of sustainable growth, the Spanish economic model had collapsed. The catalyst for this was the subprime crisis in 2007 in the US, which had important consequences all over the world, but particularly so in Spain, where private debt of both banks and households were already extremely high (Royo, 2013; Stiglitz, 2012; Krugman, 2012; Sala and Martín, 2011). The construction industry slowed down as real estate prices continually fell, which generated a continuous decline of real estate assets, most of them in the hands of Spanish banks. As a result, by the end of 2009, economic recession and the banking crisis were already monopolizing the governmental agenda.

From 2009, the economy, and especially the banking sector, has monopolized the Spanish political agenda. One of the first policy decisions was the creation of the *Fondo de Restructuración Ordenación Bancaria* (Fund for Restructuring and Reorganizing the Banking Sector, so-called FROB) in June 2009 for restructuring the financial sector.[2] Later in 2010, the *Ley de Organos Rectores de Cajas de Ahorro* (Law relating to the Ruling Bodies of Savings Banks) introduced one of the most radical changes in the financial sector designed to professionalize and depoliticize financial institutions.[3] All these measures were important to increase the solvency and strength of the financial sector, but they were not able to prevent the intervention of EU institutions in the political and economic sphere.

In September 2011, the Spanish Parliament approved a constitutional reform designed to mitigate concerns over the willingness of Spanish governments to control public finance. According to Article 135, interest and principal payments are prioritized over any other public expenditure, and from 2020, public debt of the Spanish and regional governments has a limit imposed by law.[4] This happened a few weeks after European political elites, namely German Chancellor Angela Merkel and French President Nicolas Sarkozy, highlighted the importance of imposing debt limits by law at the national level, and the European

Central Bank (ECB) started buying Spanish bonds as part of its initial rescue program.

Despite high institutional constraints and veto points, constitutional amendment took place in less than two months, as a result of the agreement of the two main political forces, which accounted for 92 percent of the seats in the Spanish Parliament. This agreement illustrates the stability of economic reforms under the PSOE and PP governments. In fact, the constitutional amendment was also part of the negotiations between Spanish and EU political elites to set the basis of the economic rescue plan for Spain; in June 2012, the Eurogroup already announced a rescue plan allocating 100 billion euros for support to the bailout of a large number of banks (Jordana, 2014; Ortega and Peñalosa 2012). These economic reforms and rescue plans had a radical impact, not only on economic and banking policy, but also on redistributive policies and welfare-related issues.

By 2013, public expenditures in Spain had been redirected to pay interest-debt payments (which increased more than 100 percent, from 17 billion euros in 2007 to 35 billion euros in 2013) to transfers of capital to cover unemployment benefits (which increased by more than 38 percent in the last five years[5] as the unemployment rate rose to more than 25 percent in 2013. For the last three years, the government of Mariano Rajoy focuses increasingly on economic and labor issues, following the mandate of EU authorities, leaving any other issue in the policy agenda in second stage. This is illustrated in Table 3.4.

Table 3.4 presents a series of simple regression models. For each of the 19 CAP topics, the table shows the impact of right-wing (PP) control of the government and the unemployment rate on the share of attention in the Prime Minister's speech. In this way, controlling for which party controls the government, we can see whether bad economic conditions lead to increased or decreased attention to any given issue. Results show that high unemployment leads to increased attention to macroeconomics, commerce and banking, and foreign trade issues, but that several other issues see reduced attention when unemployment is high. We see statistically significant reductions in attention to rights, education, environment, transportation, crime, and scientific research. These issues are simply pushed off the agendas when economic conditions are bad, as governments of both political sides focus increasingly only on the economy. This crowding-out effect of important social issues by an increased focus on economic growth has generated an intense political debate characterized by the proliferation of mass demonstrations such as the 15-M movement, or the Stop Evictions Movement (Della Porta and Mattoni, 2014). Table 3.4 makes clear that attention scarcity has

Table 3.4 The effect of unemployment and partisan control on executive attention

		Speeches			Executive bills		
Topic		**B**	**Sig.**	**R^2**	**B**	**Sig.**	**R^2**
Economy	Unemployment	1.057	.000	.534	.520	.199	.112
	Party preferences	.551	.751		4.90	.155	
Rights	Unemployment	–.646	.000	.472	–.446	.046	.209
	Party preferences	–1.217	.319		–1.630	.379	
Health	Unemployment	–.020	.658	.473	–.077	.547	.250
	Party preferences	2.022	.000		1.832	.102	
Agriculture	Unemployment	.002	.963	.078	.012	.917	.113
	Party preferences	.459	.134		1.490	.133	
Labor	Unemployment	.087	.813	.050	.051	.802	.008
	Party preferences	–3.912	.252		–.463	.791	
Education	Unemployment	–.326	.011	.225	.074	.760	.064
	Party preferences	.780	.487		–2.248	.304	
Environment	Unemployment	–.107	.007	.280	–.363	.023	.365
	Party preferences	.430	.210		–2.256	.129	
Energy	Unemployment	.005	.866	.012	.073	.616	.045
	Party preferences	.156	.572		–.917	.589	
Transportation	Unemployment	–.151	.029	.178	–.101	.716	.006
	Party preferences	–.623	.312		–.097	.966	
Crime & Justice	Unemployment	–.692	.007	.253	–.203	.458	.026
	Party preferences	1.796	.421		–1.045	.674	
Social policy	Unemployment	–.102	.220	.078	–.009	.946	.101
	Party preferences	–.735	.337		.773	.503	
Housing	Unemployment	–.119	.163	.125	–.052	.837	.029
	Party preferences	–1.139	.146		–1.019	.616	
Commerce	Unemployment	.290	.026	.166	–.603	.101	.167
	Party preferences	.259	.821		–5.969	.067	
Defense	Unemployment	.033	.799	.010	.062	.770	.066
	Party preferences	–.533	.652		–1.615	.357	
Research & development	Unemployment	–.117	.074	.169	.144	.411	.041
	Party preferences	.793	.182		.745	.580	
Foreign trade	Unemployment	.183	.001	.388	.154	.385	.191
	Party preferences	–.977	.051		.138	.877	
International affairs	Unemployment	–.141	.669	.007	.192	.483	.046
	Party preferences	–.179	.953		1.844	.365	
Government	Unemployment	.493	.170	.094	–.076	.793	.105
	Party preferences	–2.795	.394		–4.077	.106	
Public lands	Unemployment	–.199	.161	.072	.075	.607	.023
	Party preferences	–.421	.732		–.179	.882	

Note: The dependent variable is the percentage of attention devoted to the policy in question. Independent variables are party preferences (1 if the PP is governing; 0 otherwise) and unemployment (annual average unemployment rate). We tested the same model with other indicators for the economy, such as public perceptions of the economic situation, and results are similar. Models also include a constant term, not shown. B-coefficients indicate the predicted increase or decrease in attention to that topic as the unemployment rate changes, or as partisan control moves from the left to the right.

significant costs: economic crises force governments not only to pay attention to the crisis but also, because attention cannot simultaneously be given in all directions, to reduce whatever focus they might like to have on other issues, such as civil rights or the environment.

Foreign policy and defense

While political leaders of all stripes have come to focus on the economic crisis, one of the largest transformations of the symbolic agenda for the last decade is the decline of international affairs and defense issues, in particular in the period after 2004. Figure 3.2 illustrates these trends.

As Barbé (1999) emphasizes, foreign policy was one of the main concerns of Felipe González, across the entire period of his mandate (1982–1996). First, the PSOE government gave support to the decision taken by the previous government of the UCD, to become a member of NATO, contradicting one of the key elements of its electoral

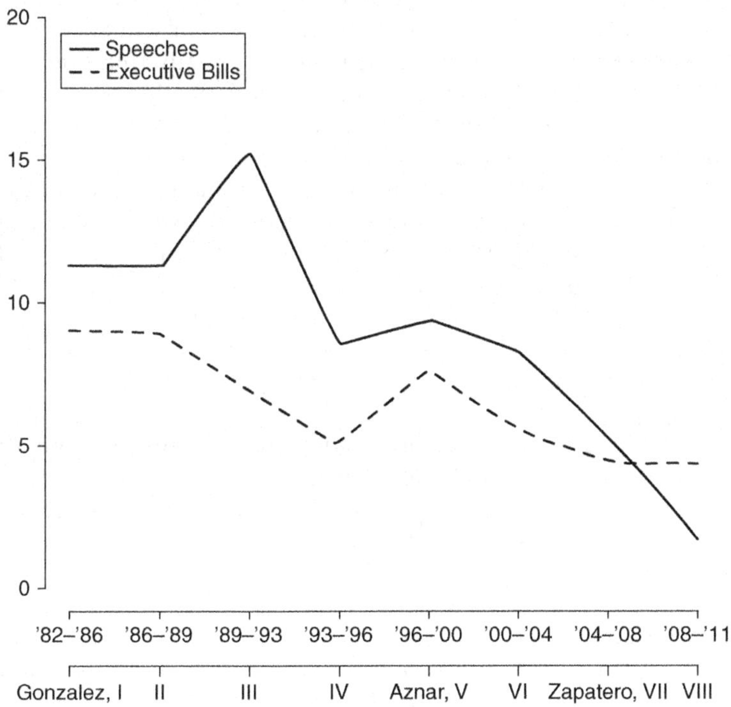

Figure 3.2 Attention to defense and international affairs

Note: Series show the percent of speeches and executive bills in each year which were devoted to either defense or international affairs.

Source: Spanish Policy Agendas Databases (www.ub.edu/spanishpolicyagendas).

program (see Maravall, 2013, for a review). At the same time, the PSOE led the entry of Spain into EEC in 1986, participated actively in the definition of the principles and goals of the Maastricht treaty in 1993, and became one of the main promoters of the Economic and Monetary Union. In this process, Felipe González was an active and legitimized policy actor who wanted to take advantage of a political context characterized by the fall of the Berlin Wall and German reunification. For more than two decades, Spain was one of the main allies of Germany and France in the process of the construction of Europe, an alliance that increased substantially the bargaining capacity of the Spanish government in the discussion of key issues like the cohesion funds (Torreblanca, 2005; Barber, 1999).However, in the early 1990s, most Spanish citizens identified European integration as one of the main causes of the economic crisis of the early 1990s (see Chapter 7). From 1996 to 2004, José María Aznar had to deal with the full integration of Spain into NATO, the professionalization of the army, the war of Iraq, the entry into the EMU, and the consequences of the EU enlargement. For some of these issues, especially those related to NATO and the EMU, Aznar maintained a strong stability regarding the policy goals of Spain with the previous government[6] (Barber, 1999). In contrast, Aznar's foreign policy differed significantly from previous periods, especially in terms of alliances with other EU leaders. The PP breaks the previous intergovernmental coalition formed by Germany, France, and Spain, in favor of the UK and US (what some called the Atlantic axis), giving support to the military intervention in Iraq, and blocking diplomatic relations in Cuba (Vilanova, 2006). Aznar becomes a close ally of US President George W. Bush and UK Prime Minister Tony Blair, with whom he also shared a common vision of economic policy.

Once Rodríguez Zapatero won election in 2004, his main priorities were to dismantle some of the key elements of the PP government's policies. This meant to call back the troops from Iraq, to normalize diplomatic relations with Cuba, and break the Atlantic axis in favor of EU counterparts. The PSOE government wanted to become a key player in the international arena, leading the so-called Alliance of civilizations.[7] It further sought to become one of the main promoters of the consolidation of the political construction of Europe, by among other things being the first country in calling for the referendum of the European Constitution in 2004. The failure of both processes (with negative results in France and the Netherlands) joined with the economic crisis, greatly limited any ambition of the Spanish executive to become a leader in the international arena.

From 2007 Spain becomes a subordinated player in the European and international arena. The EU dictates the goals and instruments of economic policy, imposing severe austerity plans, auditing financial institutions, and even forcing a reform of the Spanish Constitution in September 2011. In this new context, what was once one of the most important components of the Prime Minister's agenda and source of power is relegated to second stage. Both Rodríguez Zapatero (from 2008) and Rajoy devote less than 5 percent of the speech to deal with foreign policy, which indicates their wiliness to avoid foreign policy as a key element of the governmental agenda.

Governmental issues

Attention to governmental issues always captures most of the attention of the head of the executive in both the symbolic and decisional agendas with important variations across time, especially in the case of the PSOE governments of Felipe González. In the 1980s, attention is very much focused in the construction of governmental institutions, from the definition of the functions and structure of the public administration and executive-legislative relations, to the judiciary system and intergovernmental relations (e.g., Baena, 1997; Alcántara and Martínez 1997; Alba, 2001; Olmeda et al., 2012). Later in the early 1990s, attention is mainly related to the reform of the public sector (mostly privatization) and political scandals. As Figure 3.3 illustrates, González is the only Prime Minister who devotes a significant share of attention to corruption while giving his annual speech, among other reasons because his last two mandates were characterized by the proliferation of political scandals including the *caso Filesa* (illegal financing of the PSOE), the case of bribery and misappropriation of public resources by Luis Roldan, the general director of the *Guardia Civil* (a military police body), or the *caso Ibercorb* (fiscal fraud of Mariano Rubio, governor of the Bank of Spain), among many others (for a review, see Nieto, 1997, and Appendix B of this volume).

In contrast to González, who paid much attention to scandals, Aznar's symbolic and substantive agendas were more focused on the reform of public administration, following the goals and principles of the so-called new public management policy paradigm. The *Ley Orgánica sobre el Funcionamiento de la Administración General del Estado* (Organic Law Concerning the Functioning of Public Administration), passed in

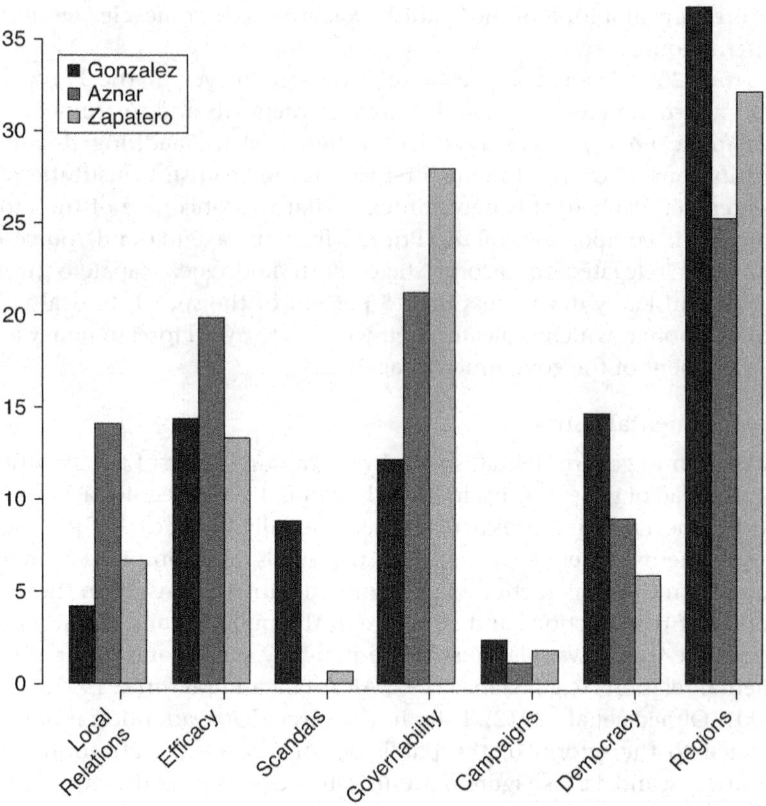

Figure 3.3 Attention to governmental affairs by three prime ministers

Note: For each Prime Minister, bars show the percent of their annual speeches devoted to the government affairs subtopics indicated. Local relations refer to mentions of state–local government relations; efficacy refers to public administration reforms; scandals refer to scandals of corruption; governability includes everything regarding the functioning of political institutions and executive-parliamentarian relations; campaigns refer to elections and political parties; democracy refers to issues related to political transition and consolidation of democracy; and regions refer to everything related to state–regions relations.

Source: Spanish Policy Agendas Databases (www.ub.edu/spanishpolicyagendas).

1997, is one of the key elements of this regulatory reform – according to which some of the legacies of the Franco regime were eliminated, like the *Gobernadores Civiles*[8] (Provincial Governors appointed by the central state). Also, the PP governments prioritized the process of privatization of public services and enterprises (Bel, 2011; Bel and Costas, 2001) and fostered the delegation of the provision of some social services to nonprofit organizations (see Pérez Diaz, 2001; Muñoz, 2013).

From 2004, attention to governmental issues increases again as a result of two factors. On the one hand, the so-called *Plan Ibarreche* and the reforms of the Estatutos de Autonomía forced the entry of state-regional relations into the agenda of Rodríguez Zapatero (see Chapter 6). On the other hand, the PSOE government prioritizes new issues in the governmental agenda, like the politics of memory (see Barahona de Brito et al., 2001; Aguilar, 1996). Transition to democracy is always an important issue in the speeches of Aznar and González, but none of them took a step further designed to define comprehensive public policies about the consequences of political violence during the Civil War and Franco's regime. Beginning in 2004, the PSOE government devotes some resources to this issue, like the creation of the National Centre for Documentation and Research on Civil War and Francoism,[9] and later in 2007 the introduction of an executive bill (finally passed as the Historic Memory Law) sought to dignify those who fought against Franco, and to compensate the victims of the civil war and the dictatorship. Mariano Rajoy is the Prime Minister that devotes less attention to governmental issues for the whole democratic period. Less than 5 percent of his first speech in 2012 is devoted to deal with these issues. As in many other advanced democracies, the government of Mariano Rajoy highlights the importance of fostering new public sector reforms as a means to respond efficiently to the crisis of governability of the late 2000s. The so-called *Ley de Transparencia* passed in 2013 as the first law in Spain aimed to deal with this issue in a comprehensive way. However, this law leaves behind many of the original promises defined in the party manifesto, like the regulation of interest groups and intervention in the governmental and parliamentary arenas among others. In contrast to previous Prime Ministers, Rajoy devotes less than 2 percent of his first speech to talk about the relations of the central government with the regions despite the importance of this issue in Spanish politics.

Terrorism

As Figure 3.4 illustrates, terrorism always captures an important share of the governmental agenda with important punctuations mainly related to terrorist attacks. In his 1988 speech, following the Hipercor terrorist attack in Barcelona in which ETA killed 22 people, Felipe González was talking about the derogation of the existing antiterrorist legislation, the introduction of terrorism as a crime in the penal code, and the failures and shortcomings of terrorist reinsertion policy.[10] In his last speech as Prime Minister in 1995, right after the killing of Fernández Ordoñez (a PP deputy of the Basque Parliament), Felipe González was denying any implication of the Spanish executive in the GAL (Antiterrorist Liberation Groups) death squads formed by public officials with the support of

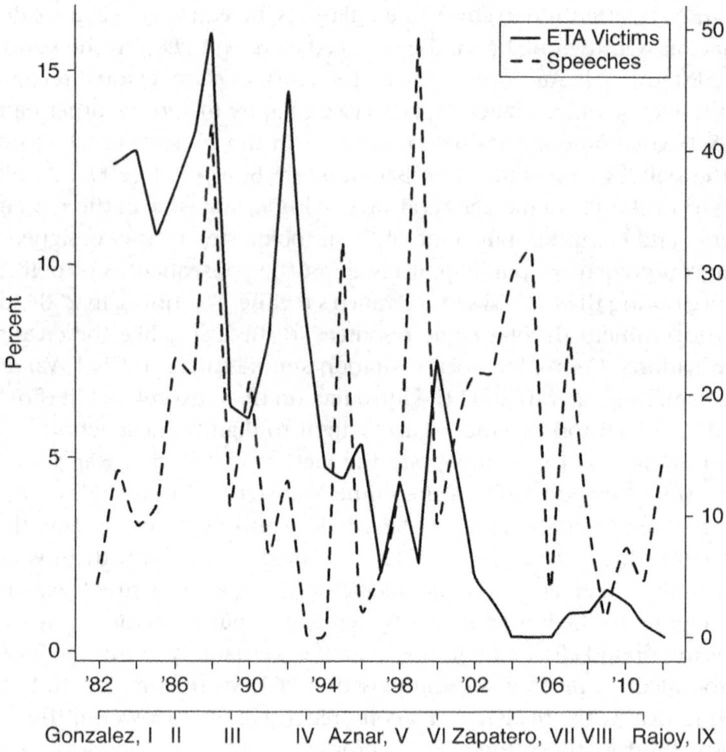

Figure 3.4 Attention to terrorism and victims of terror attacks

Note: Speeches refers to the percentage of annual Prime Minister's speech devoted to terrorism (measured on the left scale). ETA victims refers to the number of persons killed in ETA violence (measured on the right scale).

Source: Speeches: Spanish Policy Agendas Databases (www.ub.edu/spanishpolicyagendas). ETA victims: database developed by Sánchez-Cuenca and De la Calle (http://www.march.es/ceacs/proyectos/dtv/).

members of the government to fight ETA (Euskadi Ta Askatasuna), a Basque independent terrorist group – the most dramatic episodes of state terrorism in Spain since transition to democracy.

Aznar followed some of the key elements of the socialist antiterrorism policy, mainly in relation to prison policy and the intensification of cooperation with foreign countries, especially France. From the 1980s, the goal of prison policy was to guarantee the dispersion of prisoners across the Spanish territory as a means to weaken the cohesion and leadership of ETA (for a review, see Sánchez Cuenca, 2001; Reinares 2000). Aznar continues to implement these goals, but in contrast to previous years, he denies any attempt of negotiation with terrorist organizations, with a minor exception in 1998[11] and prioritizes the use of legal instruments to fight against terrorism. In this regard, the most important change was the so-called *Ley de Partidos* (Organic Law 6/2002 on Political Parties), according to which any organization that gave support to a terrorist group and/or never rejected violence as a means of politics was outlawed. Once passed in 2002, the Supreme Court outlawed several Basque political parties related to ETA – Herri Batasuna and Batasuna, Euskal Herritarrok (Basque Nationalist Action and the Communist Party of the Basque Lands) and others – but also the (reconstituted) Communist Party of Spain, because it argued that the party was part of a single structure with the terrorist group, GRAPO. The law was passed with the agreement of all political parties but the far left (IU-ICV), left regional parties including the ERC and BNG, and all Basque parties.

The PSOE shares most of the initiatives taken by the PP government. The "*Acuerdo por las libertades y contra el terrorismo*" ("Agreement against Terrorism and for Freedom")[12] signed between the PP and PSOE in December 2000, and the positive vote of the PSOE to the *Ley de Partidos* in 2002, exemplify the agreement of the two main political forces to fight together against terrorism, to intensify the dispersion of ETA prisoners, and to recognize openly their support to the victims of terrorism (about 829 killed by ETA). However, once the PSOE wins elections in 2004, Rodríguez Zapatero starts a process of negotiations with ETA, which generates an intense and highly conflictual debate in and out of Parliament. In 2005, Mariano Rajoy (the opposition leader by then) declares Rodríguez Zapatero is betraying the victims and reinvigorating ETA, emphasizing his opposition to any type of negotiation with the terrorist organization[13] (see the annual speech of 2005).

Increasing confrontation after 2004 is also linked with the Yihad terrorist attack in Madrid, on March 11, in which 191 people were killed. Despite the international dimension of this terrorist attack, the PP and some newspapers, mainly *El Mundo*, framed the issue as an internal affair that has lasted for several years (see De la Calle and Sánchez Cuenca, 2012; Reinares, 2004). Finally, attention to terrorism declines after ETA communicates officially its intention to give up violence in October 2011, one month before general elections. In this new context, Mariano Rajoy devotes only a small share of his symbolic agenda to deal with terrorism. Despite the initiative taken by ETA to start negotiations in order to find a political solution to the "Basque" issue, and the fact that we are living the longest period without ETA victims, the executive of Rajoy keeps emphasizing its strategy of no negotiation with ETA, its support to the victims of terrorism, and its policy of ignoring the "Basque" problem.

Rights

Figure 3.5 highlights the growing importance of rights-related issues – mainly immigration, moral, and gender-related issues – in the governmental agenda. Attention to immigration was almost invisible for political elites and public opinion until 2000.[14] In fact, the politicization of immigration only occurred in the late 1990s when the number of foreign-born individuals living in Spain surpassed the number of Spanish citizens living abroad. Opposition parties in collaboration with Minister of Labor Manuel Pimentel took the first initiative to change regulation, introducing a parliamentary bill[15] passed with the agreement of most

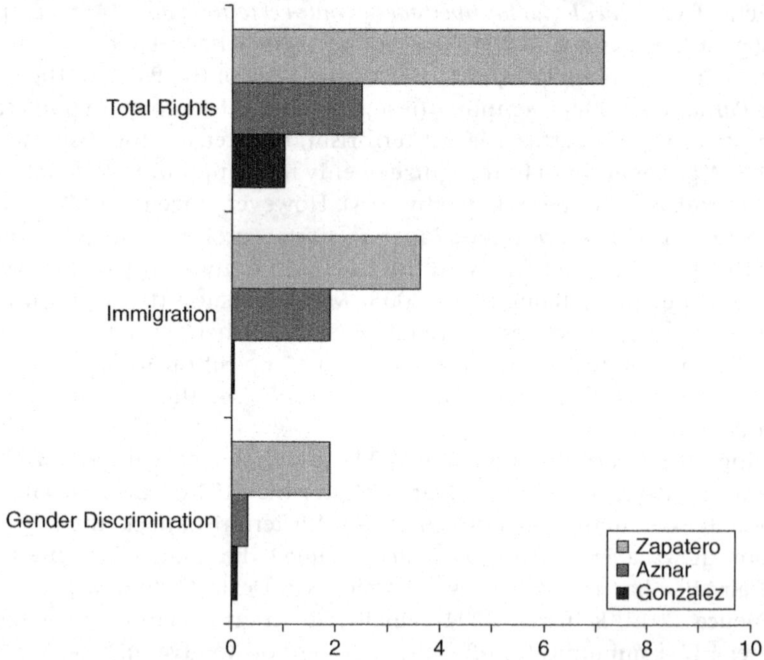

Figure 3.5 Attention to rights related issues by three prime ministers

Note: For each Prime Minister, bars show the percentage of their annual speeches devoted to rights issues. Note that immigration and gender discrimination are included in "total rights" attention.

Source: Spanish Policy Agendas Databases (www.ub.edu/spanishpolicyagendas).

parliamentary forces but the PP, a few days before general elections (for a review, see Aja, 2012). The content of this law was partially amended by two[16] new laws sponsored by the PP majority government in the next legislature, after a highly conflictual and intense political debate.

Attention to immigration increases again dramatically in the late 2000s when the number of foreign born is almost 7 million (Morales and Giuni, 2011). Rodríguez Zapatero pays increasing attention to immigration as a response of growing public concern – more than 50 percent of Spaniards considered immigration as one of the most important problems facing the nation and identified immigration as something negative for the economy and internal security (an increasing number of people identified immigration with crime), but especially EU demands. The executive bill introduced by Rodríguez Zapatero in 2008 implements EU directives regarding immigration, and at the same time, transforms the policy paradigm governing this issue for decades.[17] For the first time, immigration is defined as a labor issue, instead of a problem mainly related with public order, and immigrants' social rights are explicitly recognized (Aja, 2012).

In contrast, the entry of moral issues in the agenda is a direct response to party preferences and changing public attitudes (Chaqués-Bonafont and Palau, 2012). Rodríguez Zapatero sought to transform the regulation of moral issues by modifying previous laws passed in the 1980s – abortion, assisted reproduction technologies, and stem-cell research – and incorporating new issues in the agenda, including same-sex marriage. As we explain in Chapter 8, joined with education policy, regulatory change on moral issues generates a new context of direct confrontation with conservative forces, especially the Catholic Church, which still prevails as the source of one of the key policy cleavages in Spanish politics.

Finally, the passing of *Ley Orgánica 3/2007 para la Igualdad Efectiva de Hombres y Mujeres* (Organic Law for Effective Equality between Men and Women) in 2007, jointly with *Ley de Dependencia* (see next section) is one of the milestones of the PSOE government's second legislature. The passing of this law is designed to deal with gender discrimination in a comprehensive way, addressing issues like violence against women (about 70 women are killed annually by their partners or ex-partners; see www.msssi.gob.es); salary inequalities between men and women[18] (women's salaries are 20 percent lower in Spain than men's); discrimination against widows (widows' pensions are under the poverty line); and also guaranteeing access of women to positions of the highest responsibility in economic and political institutions.

Still, the implementation of these policy goals has been relegated to second stage as a result of the economic crisis and the lack of political will of governmental actors, most of whom are, of course, men. In his first annual speech, Mariano Rajoy devotes just a couple of sentences to deal with rights-related issues, despite one of the main policy promises of the PP being to revoke most policy decisions related to moral issues taken from 2004 under PSOE governments. The PP sent to the Constitutional Court a revision of the organic law that regulated same-sex marriage and the one that regulated abortion. Once in office, despite having the absolute majority of seats, rights issues are off the agenda. As we explain in Chapter 8, the PP introduces a bill to reform the abortion legislation passed by Rodríguez Zapatero, but the project was finally shelved, mainly as a result of changing values of Spanish society and electoral strategies.

Welfare

Attention to welfare-related issues varies significantly across time. In the 1980s, the PSOE governments became the main builders of what Esping-Andersen (2000) categorizes as a Mediterranean corporatist welfare state, characterized by access to all citizens to welfare benefits (universalism); the predominant role of the state as the institution that regulates, implements, and finances welfare policies (statism), inequalities – in terms of the generosity of coverage depending on the sector of activity (corporatism) – and strong dependency on families for the provision of an important share of social services like child care, or elderly care (familiarism). Actually, Spanish public expenditure grew from 20 percent of GDP in the mid-1970s to almost 50 percent by 1993, almost reaching the average public expenditure in social services of EU countries (Boix, 2006).

The PSOE governments sought to go beyond most of the reforms made by the Unión de Centro Democrático (UCD) governments during transition to democracy, and more specifically to reverse the fragmented, uncoordinated, and inefficient social insurance schemes developed during Francoism, reducing at the same time, the control of the Catholic Church over morality and education (Moreno, 2009). To do that, the PSOE governments sponsored two major laws. On the one hand, *Ley Orgánica de Derecho a la Educación* (LODE 8/1985, Organic Law about the Right to Education) recognized the obligation of the state to finance compulsory education on the basis of a dual system where private institutions, overwhelmingly religious, were entitled to be subsidized by the state (Magone, 2004). In 1990, *Ley Orgánica General del Sistema Educativo* (LOGSE 1/1990, Organic Law of the Education System) extended compulsory education to 16 years avoiding the early division of students between academic and vocational streams, one of the main causes of academic inequalities in Spain (Bonal, 1998; Guillén and León, 2011).

These reforms were designed to guarantee equality of access to education and were accompanied with an exponential increase of public expenditures on education – from 1975 to the mid-1990s public expenditure on education increased from 2.2 to 4.5 percent of the GDP, and the number of scholarship holders increased by 4.7 percent (Bonal, 1998). On the other hand, *Ley General de Sanidad* (14/1986, General Law on Health), passed in 1986, created the conditions to consolidate a national health service, characterized by the universal access (implemented in full by 1989) and the public provision of health services (Moreno, 2009; Magone, 2001; Guillén and León, 2011; Subirats and Gallego, 2002; del Pino, 2013).

From 1996, the PP governments promoted the privatization of public services, especially in relation to health issues.[19] Private contributions to total health expenditure increased from 21.3 percent in 1990 to 30 percent in 2005 (Rico and Sabés, 2000), social services were increasingly outsourced, and overall levels of social protection were reduced. One of the few areas in which public expenditure increased was family policy.[20] In this regard, Aznar's government did not seek to transform the strongly family-focused[21] characteristics of the Spanish welfare state that have been apparent for many decades, but rather to counter decreasing fertility rates, one of the lowest in the world, while also responding to EU regulations on family issues at the same time (Salido, 2011; Bianculli and Jordana, 2013).

The conservative government also passed major reforms on education. With the *Ley Orgánica de Calidad de Educación* (LOCE 10/2002, Organic Law about Education Quality), Aznar intended to halt further steps toward reducing the Church's role in education and reinstated practices that could advance the Catholic cause by providing new advantages for the Church like funds from public coffers, tax exemptions, the right to appoint and dismiss religious teachers in public schools, and reestablishing religious education in all public schools. The reform was also designed to reestablish the division of students between two different streams – the academic and the vocationally trained – which, in practice, was meant to divide students according to their academic results.

Overall, this reform was strongly criticized by most opposition parties, among other things for being discriminatory and for undermining equality of opportunities. Accordingly, following one of the PSOE electoral promises, Rodríguez Zapatero passed a new law, *Ley Orgánica de*

Educación (LOE 2/2006, Organic Law of Education), in 2006, making religious education noncompulsory and introducing the controversial subject "Citizenship Education," which included issues like single parenting and gay families in the curriculum (Magone, 2001). The Catholic Church directly promoted major demonstrations in collaboration with civil society organizations like the *Confederación Católica de Padres de Família y Padres de Alumnos,* and with the support of conservative parties, mainly the PP. In the same vein, the PSOE government also revoked some of the content of *Ley Orgánica de Universidades* (*Ley 6/2001*, Organic Law of Universities), passed in 2001, with a new version (*Ley 4/2007*) in 2007, which sought to enhance the autonomy of universities and to strengthen the relationship between the private sector and the academic world as part of the wider plan of increasing the level of investment in R&D.

But, the milestone of Rodríguez Zapatero was *Ley 39/2006, de 14 de diciembre, de Promoción de la Autonomía Personal y Atención a las personas en situación de dependencia* (Law for the Promotion of Personal Freedom and Care for Those in a Position of Dependency). This law was designed to create a universal long-term care system covering the needs of the dependent population and to put an end to the dilemma faced by many Spanish women – either to have children or get a full-time job. In fact, the goal was to transform welfare provision into something more similar to the social-democratic model (characteristic of Scandinavian countries) in which household welfare and caring responsibilities are relaxed through state provision (that is, the policies are focused on reducing the reliance on family, especially women, for the provision of widely needed social welfare benefits). In this regard, the PSOE also introduced reforms to balance work and family life (*Ley Orgánica 3/2007, para la igualdad efectiva entre hombres y mujeres,* Law for Effective Equality between Men and Women). However, social-protection spending, especially designed to support families, the disabled, and the aging population, still lags behind most EU countries (Moreno, 2009; Navarro, 2009). This is especially the case after the economic and financial crisis. Rodríguez Zapatero had to implement severe cutbacks in social welfare programs and major labor-market reforms, failing to fulfill an important part of his electoral promises.

Mariano Rajoy's conservative government continued to implement policies to curb spending and control public deficit, launched a new labor-market reform, and raised taxes, contradicting some of the PP ideological principles and electoral commitments. By the same token, Rajoy's government sponsored a set of policies designed to revoke most of the welfare reforms sponsored by the PSOE. The so-called Wert

Law (*Ley Orgánica 8/2013, para la Mejora de la Calidad de la Enseñanza,* Improving the Quality of Education Law, LOMCE) introduced significant adjustments into the curriculum following a strong recentralization approach. Formally designed to reduce high dropout rates and low achievement levels, the law incorporates controversial aspects such as a bigger role for religion and state funding for schools that segregate students by gender. In short, this reform illustrates the intensification of politics of resentment and lack of consensus, in which governing parties devote an important part of their efforts to revoke the decisions of previous governments. The same occurs in other policy areas, especially the environment.

Environment

Environmental issues were not a top priority for the PSOE governments of Felipe González – less than 1 percent of his speeches and 4 percent of the bills. For the first decades of Spanish democracy, environmental policy was characterized by a reactive and uncoordinated approach, mainly focused on the protection of natural spaces and water policy, and very much connected with EU decisions (only in 1995 did attention to environmental issues increase more than 13 percent as a result of the implementation of EU directives). In this regard, *Ley de Aguas* (*29/1985,* Water Law, sponsored by the executive and passed in 1985) was designed to regulate industrial waste in view of the entry into the EEC, and especially to overcome the water deficit problem (Aguilar, 1997: 258). The severe droughts in the 1990s made clear the failure of previous policies and the need to implement a new *Plan Hidrológico Nacional* (PHN). The last government of Felipe González introduced a bill in 1995 to fulfill this goal, but the economic costs of the project, the concerns raised by water-rich regions from the north (questioning the invocation of national solidarity), and the lack of an alternative innovative approach to hydraulic policy led to the defeat of the PSOE proposal (Bakker, 2002; Arrojo, 2005; Pérez Díaz, 1996).

The PP led the first government, which created a Ministry of Environment and a new Advisory Council on the Environment (*Consejo Asesor de Medio Ambiente*), including representatives of interest groups and national and regional authorities, in 1996 (Prindham and Magone, 2006). However, most of the PP policy decisions regarding environmental policy are directly linked with EU imperatives. The adaptation of domestic legislation to European standards was hampered by the low development of environmental regulation previous to accession, the high costs associated with implementation of EU directives, and the lack of party consensus regarding high standards of environmental protection in relation to air pollution, waste management, and disposal of waste water (Aguilar, 1997; Börzel, 2000).

Increasing attention to environmental issues by the PP governments is also linked to changing public attitudes, dramatic events related to environmental catastrophes (e.g., the sinking of the Prestige in 2002), and the controversy surrounding some policy decisions (e.g., water policy). In 2001, the conservatives passed a new law (*Ley 10/2001, del* PHN) designed, among other elements, to pipe water from the River Ebro down to Southern Spain, generating an intense and highly controversial debate in and out of the Spanish Parliament. EU institutions questioned the compatibility of this new plan with the new framework directive on water passed in 2000, civil society groups along with some regional political elites complained about the costs (economic but also in terms of cultural heritage) of the implementation of this new water policy for some regions, while experts claimed the need to change water policy toward the adoption of a new policy paradigm based more on water governability and the adoption of a new culture of water (see Arrojo, 2005, for a review).

This "new culture of water" was promoted by a vast set of interest organizations and social movements that emerged alongside an increasing environmental awareness of Spanish citizens. This is illustrated by the increase of membership in organizations like Greenpeace Spain, which grew from 16,000 members in 1984 to more than 85,000 in 2004, or by the rise in the environmental associations registered with the Ministry of Environment (Magone, 2004). From 2004, the PSOE governments fostered policy change toward this new culture of water and governability of water resources, cancelling the most controversial aspects of the previous PHN. Actually, Rodríguez Zapatero devoted about 7 percent of the policy agenda to deal with environmental issues, fostering some highly controversial policy decisions like the Coastal Law, which among other things prohibited the privatization of the sea coast. One of the only policy decisions taken by the PP government of Mariano Rajoy regarding environmental issues has been to revoke most of these policy decisions by way of passing a new Coastal Law in 2013, which basically reduces the levels of coastal protection in favor of private economic activities.

In sum, for the last 15 years there has been a highly controversial dispute about environmental issues, in which the PP clearly prioritizes economic interest, and the PSOE (not so clearly) tries to establish an equilibrium between economic interest and environmental protection, especially from 2004 to 2008. These disputes on environmental issues also illustrate the politics of confrontation that characterize Spanish politics in past decades. As in the case of education, moral issues, foreign policy, or immigration, both the PP and PSOE devote their resources to revoke the policy decisions taken by previous governments, denying more and more the role of Parliament as a space of political debate. We review these trends toward greater political polarization and the transformation of the role of Parliament in the next section.

The legislative agenda-setting capacity of the executive

To achieve the various goals outlined above, Spanish governments – from González (1982–1996) to Aznar (1996–2004), Rodríguez Zapatero (2004–2011), and Rajoy (2011–2013) – have introduced an average of 45 governmental bills and 13 decree-laws a year, of which 92 percent and 98 percent, respectively, are passed into law. This is, in contrast to other advanced democracies (see Döring, 1995: 42) in Spain, the executive is the main producer of legislative decisions, with a great power to translate policy promises into policy decisions, even within Parliament. In fact, more than 90 percent of the laws passed from 1982 to 2013 – 1172 ordinary laws, 281 organic laws, 418 decree-laws, and 55 legislative decrees – have been elaborated and promoted by the governing party, with some interesting variations across time.[22]

The production of governmental bills follows a cyclical pattern (Figure 3.6). All Prime Ministers try to legislate more during their first mandate, and this is especially the case for González (from 1982 to 1986, the PSOE government introduces 33 percent more bills than for the rest of legislatures) and José María Aznar (the PP government introduces 10 percent more bills in the first mandate), while in the case of Rodríguez Zapatero the number of governmental bills are only slightly larger during the first mandate. Also, as in other countries (for the US, see Jones and Baumgartner, 2005), bill production is larger during the second and third years along the legislature, declining dramatically during electoral years without variation across governments.

In contrast, the production of decree-laws follows an incremental trend (Figure 3.7). More than 50 percent of the laws passed from December 2011 to September 2014 are decree-laws, versus an average of 16 percent during González's PSOE governments, 22 percent during Aznar's, and 25 percent during Rodríguez Zapatero's. Also, less than 9 percent of the decree-laws introduced by Rajoy are related to urgent needs like environmental catastrophes or unexpected events (e.g., earthquakes, droughts

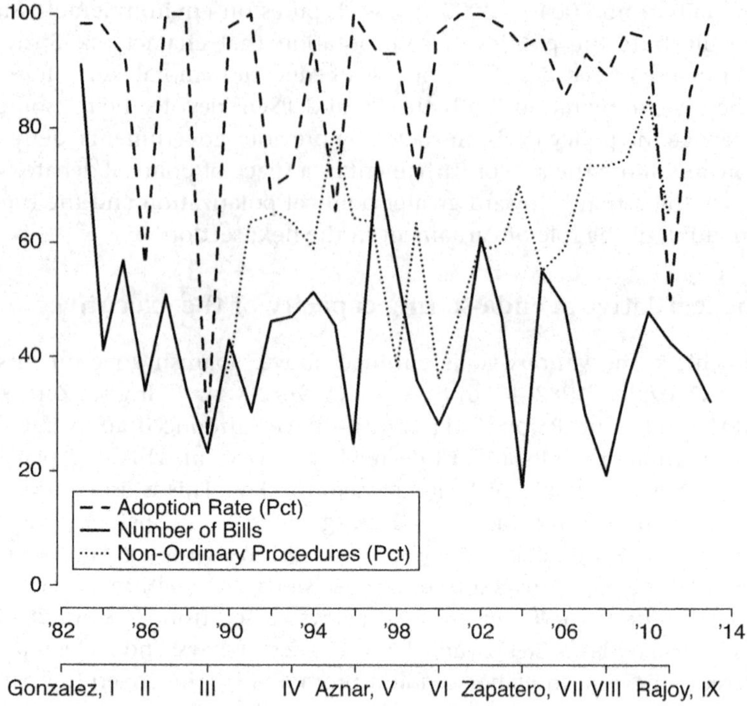

Figure 3.6 Legislative agenda-setting by the executive

Note: Series show the number of bills per year, the percentage of bills adopted, and the percentage of bills subjected to non-ordinary procedures.

Source: Spanish Policy Agendas Databases (www.ub.edu/spanishpolicyagendas).

or fires). On average, these unexpected events account for 45 percent of the decree-laws passed under González's governments, 60 percent in the case of Aznar, and only 25 percent in the case of Rodríguez Zapatero. That is, the Spanish executive is increasingly using decree-laws as the main legislative instrument to make decisions about a vast array of issues that have nothing to do with an urgent necessity. This has important consequences for the functioning of democracy, and especially on the role of Parliament in the decision-making process. Decree-laws by definition deny MPs the capacity to introduce amendments, neglect the importance of parliamentary debate, and contribute to increasing polarization and confrontation in Spanish politics.[23]

In the same vein, most executive bills are increasingly passed using non-ordinary procedures – urgent and commissioned procedures – about

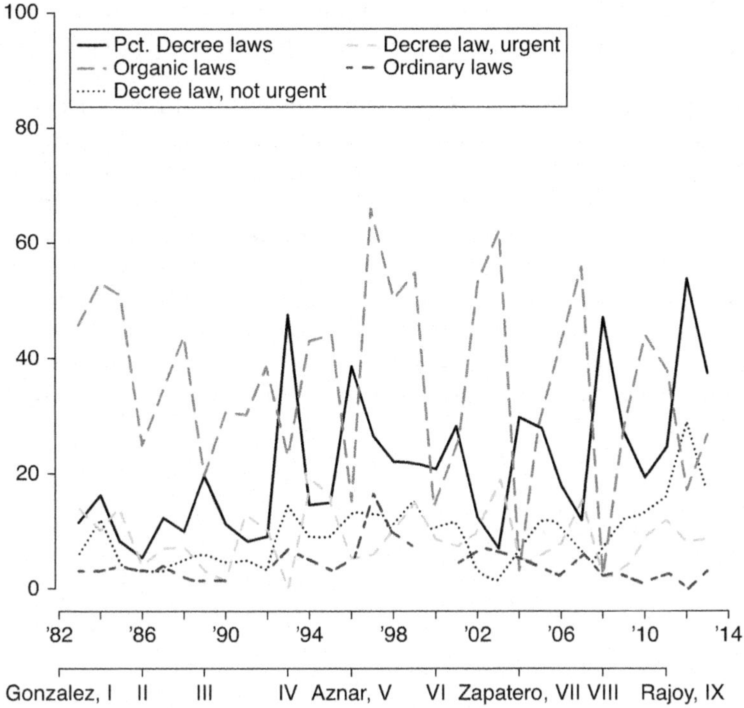

Figure 3.7 Evolution of different types of laws

Note: Series show the number of each type of law, with the dark solid line showing decree-laws as a percentage of all laws.

Source: Spanish Policy Agendas Databases (www.ub.edu/spanishpolicyagendas).

80 percent from 2008 to 2011, which basically means they were not discussed in any plenary sessions of the *Congreso de los Diputados*, versus an average of 55 percent for the rest of legislatures. The use of non-ordinary procedures has been traditionally understood as a strategy to promote MP specialization across policy areas, and/or to increase the capacity of Parliaments to analyze and discuss governmental legislative proposals (e.g., Döring, 1995). In a context of growing government intervention, the use of non-ordinary procedures may be a rational and efficient instrument for MPs to deal with a growing number of governmental bills, avoiding overloads in the parliamentary arena. But as Figure 3.6 illustrates, the number of bills has not increased in the last years, and even more interesting, the increasing use of non-ordinary procedures does not imply a decrease in the time needed to pass bills

into laws. On the contrary, the average number of days needed to pass a bill into law increases from 1983 to 2011 with the only exception being social policy.

Faced with the power of the executive, members of the parliamentary opposition counteract with increasing confrontation and lack of support to most governmental initiatives (see also Chapter 4). The Spanish executive finds less and less support from other political forces to pass its legislative proposals, especially when governing with the majority of seats (Figure 3.8). That is, the number of MPs giving a positive vote in the final voting of organic laws or decree-laws declines gradually from 2000 to present (Palau, Muñoz and Chaqués-Bonafont, 2014), illustrating the lack of consensus and the unwillingness of parliamentary groups to reach an agreement regarding fundamental issues, especially when the executive is governing

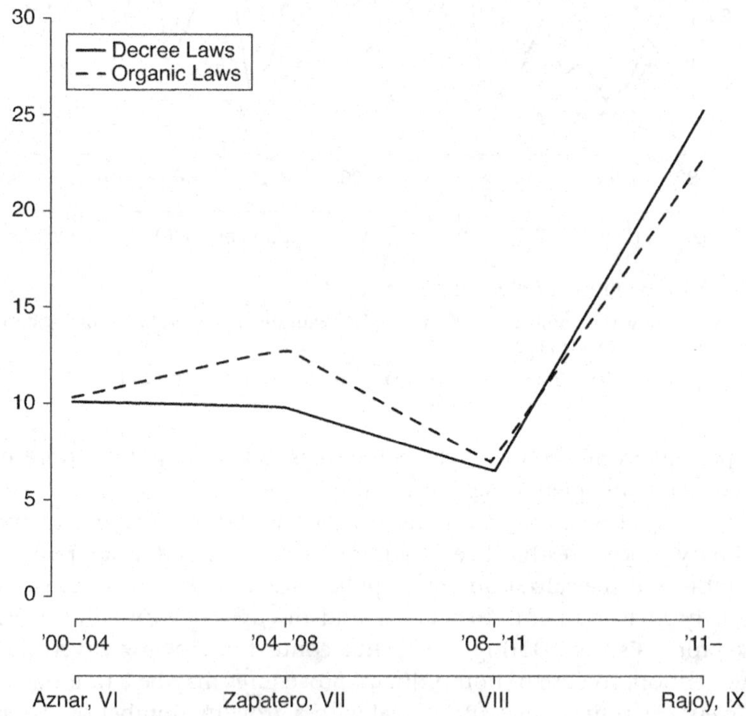

Figure 3.8 Mean percentage of negative votes for legislation

Note: The figure shows the percent of parliamentary votes against the two types of laws. Data are organized by legislature.

Source: Spanish Policy Agendas Databases (www.ub.edu/spanishpolicyagendas).

Table 3.5 The impact of minority government status on the adoption rate of executive bills

	B	Sig.	R^2
Adoption Rate			.02
Constant	87.989	.000	
Minority government	–6.284	.448	

Note: The model shows an OLS regression predicting the adoption rate of executive bills (number of bills passed/total introduced) by minority party status of the government (a dummy variable scored 1 during minority governments and 0 when the executive is governing with an absolute majority of seats). Results are non-significant. Data are annual from 1983 through 2013.

with the majority of seats.[24] The same thing occurs regarding the adoption rate of governmental bills. Results of a simple regression model confirms that the adoption rate tends to decline when the executive is governing under minority, although the coefficient is not statistically significant and the robustness of the model is very low, as shown in Table 3.5.

Table 3.5 shows that variations in the adoption rate of governmental bills cannot be explained simply by taking into account the type of government. One plausible explanation would be the idea that under minority, governments would drop off from their agendas those issues that are more controversial, prioritizing those for which they know they can find support in the Parliament. Such a strategy of avoidance could explain why the adoption rates do not vary much from minority or majority governments, as a party governing under minority might avoid the introduction of bills that have more chances to be blocked by opposition parties. If that argument is correct, it would also imply that correspondence between electoral promises and policy decisions would decline especially when governments do not enjoy the majority of seats in the Spanish Parliament. That is, mandate responsiveness would go up and down depending on institutional factors, mainly the type of government. In the next section, we demonstrate that this is not a complete explanation, at least for the case of Spain.

Broken promises

According to the party mandate theory (Budge and Hofferbert, 1990; Klingemann et al., 1994), we should expect political parties during the electoral campaign to clearly announce their policy priorities for the next legislature, and once in government they try to fulfill their mandate as a means to gain reelection, and avoid the political costs associated with the lack of implementation of policy promises (Maravall, 1997, 1999, 2013; Chaqués-Bonafont et al., 2014; Gunther and Montero, 2009). In this view, the ability of the incumbent government to fulfill promises depends on institutional factors, mainly whether there is a majority government or not, leaving other explanatory variables further down the list of priorities, especially those related to preferences and flows of information (Baumgartner et al., 2011). That is, under minority, the governing party leaves aside those bills that are more controversial and have lower probabilities to achieve a positive vote in the Congreso de los Diputados, like the case of abortion when the PSOE depends on regional conservative parties to govern (see Chapter 8), or regional fiscal reform during the first legislature of the PP (see Chapter 7).

This argument contradicts the policy dynamics approach. As Baumgartner and Jones (1993, 2015) emphasize, the policy process is an essentially disorderly and unplanned process in which party preferences and institutional factors are but one of the variables to be taken into account to explain policy outcomes. Issue prioritization varies across a legislature in response to external events and/or new ideas and information, which were not necessarily taken into account by the incumbent government when defining the party manifestos (Baumgartner et al., 2011). Parties, and political elites in general, are unable to fully predict which issues should be prioritized along the term, and this is due to cognitive and institutional constraints (see Chapter 1). As a result, the governing party may not fulfill its electoral promises (see Chaqués-Bonafont et al., 2014). It may make efforts to do so, but economic, political, and other events can conspire to require attention to topics that need addressing, even if they were not part of the electoral platform. Some items of bad news, after all, are never likely to be part of a party's platform for election, but they have to be dealt with nonetheless once a party is in government.

To test these assumptions we run a binary logistic regression model in which the dependent variable is what we defined as "broken promises," measured as those issues mentioned in the party manifesto, but not enacted by the governing party during the subsequent legislature. We created this variable at the subtopic level. All issues (subtopics in the CAP classification) mentioned in the party manifesto but not having a governmental bill are identified as 1; all issues mentioned in the party manifesto and having an equivalent governmental bill are defined as 0 (in these cases, the government did indeed enact a bill on a topic it said it would). As an example, along the first mandate of Felipe González, we identified 72 issues (28 percent of all subtopics, according to CAP methodology; see Appendix A) mentioned in the party manifesto but without a response in the decisional agenda (governmental bills introduced); during the second mandate of Aznar about 83 issues (33 percent of all subtopics) had no response; while in the last mandate of Rodríguez Zapatero it was 136 issues (54 percent of all subtopics) without decisional response, as Figure 3.9 shows. Note that there is always a set of issues that are mentioned in the party manifesto, but whose response does not necessarily require a new bill. This is the case of international relations, for example. To enhance cooperation in cultural and commercial terms with Latin America or Mediterranean countries does not necessarily require the passing of a new bill.

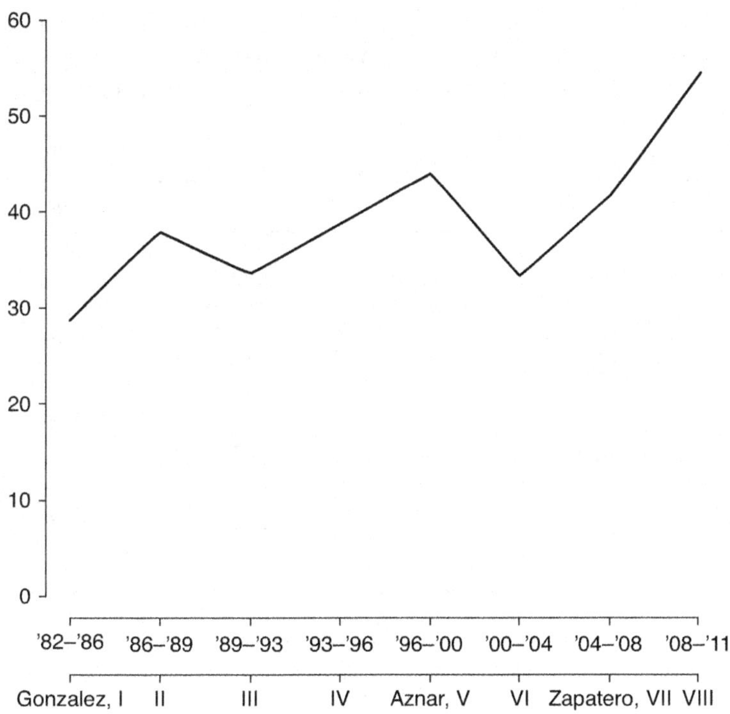

Figure 3.9 Broken promises

Note: For each legislature, the figure shows the number of times that an issue (measured by subtopics) is mentioned in the party manifesto but is not the object of a single executive bill during the legislature.

Source: Spanish Policy Agendas Databases (www.ub.edu/spanishpolicyagendas).

The independent variables are (1) the type of government (measured by a dummy variable: 1 when the party is governing with the majority of seats, and 0 when it is governing under minority) and (2) the public mood (which measures variations in public perceptions about the economy and the political situation; see Chapter 2 for a full description). Results are summarized in Table 3.6. Overall (i.e., for the period of 1982–2011), coefficients indicate that broken promises increase during economic crisis and when the governing party does not have the majority of seats in Parliament. However, results are only significant for the case of type of government. At first glance this is also illustrated by Figure 3.9, but especially for the case of the PP. Mandate responsiveness declines especially when Aznar and Rodríguez Zapatero are governing with the minority of seats. Clearly, different patterns apply to the different periods.

When we run the analysis separately for each of the Prime Ministers, results clearly indicate that in the case of Felipe González broken promises are explained by changes in the economic context more than the

Table 3.6 Broken promises, the economy, and government status

	B	Sig.	R^2
Total			.006
Type of government	.799	.069	
Economy	1.083	.517	
González			.006
Type of government	.838	.420	
Economy	1.506	.078	
Aznar			.009
Type of government	.671	.088	
Economy*			
Zapatero			.007
Type of government*			
Economy	.694	.122	

Note: The table shows the result of a logistic regression where the dependent variable is "broken promises" (scored 1 for those subtopics mentioned in the party manifesto but not having a governmental bill passed during the following legislature). The independent variables are the type of government (dummy variable with value 1 during minority governments and 0 when the executive is governing with the absolute majority of seats), variations in the perception of the economic situation (dummy variable with value 1 when the percentage of citizens considering that the economic situation is bad or very bad is above the mean of the period). *Zapatero never ruled under majority government, so that variable is excluded; and Aznar never ruled during a period of extremely negative perceptions of the economic situation, so that variable is excluded. The equations also included constant terms, which are not shown. Data cover the period of 1982 to 2011.

type of government (results are only significant for the case of the economy). This gives support to previous analyses by Stokes (1999, 2001) and Przeworski (1991; Prezeworski et al., 1999) about the impact of the economy upon mandate responsiveness. Actually, as Maravall (1999) illustrates, it is during the first mandate of Felipe González, when the PSOE achieves the absolute majority of seats in the Spanish Parliament, that broken promises are especially evident in relation to issues like NATO, labor-market reforms, and industrial policy.

In contrast, in the case of Aznar, results give some support to the importance of the type of government. In this case, broken promises are larger when the PP is governing under minority. Note that in this case, the model only considers the type of government as an explanatory variable, because as we explained before (see Figure 3.1) the economic situation from 1996 to mid-2000 was very positive (so it is impossible to control for a possible effect of a bad economic situation). In contrast, insofar as Rodríguez Zapatero is always governing under minority, the model in that case only considers economic factors as an explanatory variable. In that case, positive and significant coefficients indicate broken promises increase as the economic situation declines.

In sum, results so far give partial support to both mandate theory and the policy dynamics approach. Examples of each can be found. Economic crises in the early 1990s and late 2000s have a direct impact on the governmental agenda of Felipe González and Rodríguez Zapatero. In both cases, changes in the economy alter governmental priorities, mainly focused on the problem of unemployment, labor reforms, and the problems derived from increasing public deficit, leaving aside some of the issues that were highlighted during the electoral campaign. Also, the type of government is an important variable explaining why broken promises are larger during the first mandate of Aznar. His alliance with regional parties, formalized in the *Pacto del Majestic*, impose important variations in its initial electoral program. However, the type of government is not a significant variable to explain variations in broken promises in González's governments. Next, we analyze whether broken promises are larger for those issues with shared jurisdiction.

Multilevel governance and broken promises

Following some of the ideas already discussed in Chapter 1, we should expect broken promises increase with the consolidation of a multilevel system of government, and especially for those issues with shared jurisdiction. This is so for several reasons. First, despite issue, jurisdiction has been dramatically transformed for the last decades (see Chapters 6 and 7), as political parties keep paying attention to issues taking into account the electoral rewards associated to their politicization. That is, parties tend to politicize the issues "they own" independently, as these issues are increasingly Europeanized or regionalized. Large political parties like the PP and PSOE would not renounce the benefits associated with the politicization of issues like the construction of high-speed trains, and more specifically the *corredor central* (a high-speed train connecting Spain with France with the construction of a tunnel that crosses the center of the Pyrenees) instead of the *corredor mediterráneo*, simply because the EU has the final decision about its construction. Actually, the EU already stated its total opposition to this infrastructure, mainly because of the economic and environmental cost that it would represent for European citizens (Bel, 2011), but both the PP and PSOE have campaigned for the last years, highlighting the benefits of the *corredor central* versus any other alternative to connect Spain with the rest of Europe.

Second, the costs of the lack of implementation of policy promises are especially low in a multilevel system of governance, where the attribution of responsibilities is less clearly defined and citizens have a less precise idea of which government is politically accountable for each policy area (Chaqués-Bonafont, Palau and Muñoz, 2014; Lago and Lago, 2011). The allocation of responsibility for policy outcomes is increasingly complex, generating important incentives to political parties to engage in a process of blame-shifting and credit-taking depending on whether the news is bad or good (Anderson, 2006; Soroka and Wlezien, 2010). In this context, political parties keep paying attention to the same issues in their party manifestos, knowing that in case of inaction they can always blame the EU or the regions. Statewide and regional parties with radically different ideological positions would blame the EU, and more specifically, the economic adjustment policies imposed by the European Commission regarding the reduction of public expenditure on welfare provision.

Finally, in a multilevel system of government, the capacity of governmental actors to formulate and implement policy decisions decreases as a result of increasing institutional friction and transaction costs. Water policy, and the so-called PHN, was one of the main electoral promises of the PP during the electoral campaign of 2000, but it was never put into practice as a result of the opposition of the EU (which denied the financial resources needed to implement this policy), regional governments, and social movements, which joined with EU policy actors mobilized against the PP proposal. The same occurred with the Coastal Law of 2004

Table 3.7 Broken promises and shared jurisdiction

	B	Sig.	R^2
Broken Promises			.01
Jurisdiction	2.526	.000	

Note: The dependent variable is the same as in Table 3.7, and the model here includes just a constant term (not shown) and a measure of shared jurisdiction between the central government and the regions (scored 1 for those subtopics with shared jurisdiction and 0 otherwise). Data cover the period of 2000 through 2011.

sponsored by the PSOE, partially revoked by a decision of the European Court of Justice, or some of the measures adopted by the Spanish governments in the 1990s, to cut public expenditures on medicines.

In short, in a multilevel system of governance, parties may be less capable (because they lose political autonomy) and/or less willing (because they find electoral rewards) to implement their mandate, and this is especially so for those issues with shared jurisdiction. To test this assumption, we ran a binary logistic regression in which the dependent variable is "broken promises" (see previous section for a description) and the independent variable is issue jurisdiction, measured as a dummy variable (1 if jurisdiction is shared with the EU and/or the regions, and 0 if not shared). Note that the analysis was run at the subtopic level, which as we explain in Chapter 2, means that we take into account 247 different issues. This is especially important to identify whether jurisdiction is shared or not (see Chaqués-Bonafont and Palau, 2011a, for a previous discussion). Results (shown in Table 3.7) corroborate our initial expectation. Broken promises are significantly larger for those issues with shared jurisdiction (e.g., environmental issues or social policy) than without shared jurisdiction (foreign relations, defense, or civil rights).

Conclusions

Political leaders and their parties struggle to gain control of government for many reasons. One is the pursuit of power and the prestige associated with it. Another is the policy goals that motivate them. Analysis here has shown that whether the Spanish Prime Minister comes from the Left or the Right has only a mixed and partial effect on their policy focus. Examples abound – such as Aznar's association with the Atlantic axis in foreign policy, only to be reversed when replaced by a left government after the next election – where partisanship determines policy.

But cases where leadership responsibilities or institutional constraints force Spanish political leaders to deal with the policy challenges facing them in ways that cannot be strongly discerned by their partisanship are more common. Unpredicted crises, natural disasters, and unpleasant events such as economic recessions force governments to take up issues they might rather avoid, or never prioritized before their election. Social movements, political demands by outsiders, or the normal dynamics of journalistic coverage of the political process are another source of unpredictability in the political agendas of governments of the Left and Right. Institutional constraints abound, of course: shared control with the regions or the EU over certain policy topics, or minority control and dependence on allied, but really rival, parties in Parliament have a discernible effect on the ability of a government to keep its promises.

This chapter has explored the abilities of Spanish Prime Ministers to keep their promises and the reasons why they sometimes break them. We have demonstrated that correspondence between policy promises and policy decisions decline across time, especially when parties are governing under minority, in a context of economic crisis, and especially for those issues with shared jurisdiction. Also, we demonstrate governmental actors have transformed their way of doing politics toward a new model characterized by increasing confrontation and lack of consensus. Governmental actors more and more are governing by decree, neglecting the parliamentary arena, and MPs counteract by giving less and less support to any of the governmental proposals. In the next chapter, we explore legislative-executive relations in more detail by analyzing the oversight function of Parliaments.

4
The Content of the Government-Opposition Game

Since 2000, the politics of consensus that traditionally characterized Spanish politics has been replaced by the politics of resentment (Gunther and Montero, 2009). Consensus and mutual respect, symbols of the successful transition to democracy, have given way to increasing polarization of the political debate and the reactivation of old political cleavages. Departing from the traditional role as catch-all parties for the center-left and center-right, the PSOE and PP have turned to a strategy of direct confrontation (Montero et al., 2006; Vallespín, 2012). For moderate regional political parties, this turn has meant a diminishing role in government formation; they are no longer the pivotal players.

Shifting party preferences among the electorate partly explain the radicalization of the right during Aznar's second government and the leftward turn of the Rodríguez Zapatero government (Gunther and Montero, 2009). Both parties, the PP and PSOE, have enacted controversial policies designed to reverse previous governmental decisions. In 2000, Aznar reduced state social programs, sent troops to Iraq, downplayed environmental problems, restricted stem-cell research, and reintroduced religious instruction into the public school curriculum. Four years later, Rodríguez Zapatero brought home the troops from Iraq, promoted a radical change on the regulation of moral issues (legalizing same-sex marriage and relaxing abortion regulations), questioned the role of the Catholic Church in education, and increased public spending on social policy. Since 2011, the PP government of Mariano Rajoy has been following, and even intensifying, this strategy of direct confrontation, undoing some of the major socialist reforms, especially those related to moral issues.

The Spanish Parliament provides one of the main venues hosting this confrontation between political forces (Capo, 1994, 2003; López Nieto, 2001; Martínez, 2000). However, we know little how increasing polarization affects parliamentary activities and the content of the agenda. More pointedly, we do not know how the government-opposition game is now played. This chapter is designed to fill this gap. Our goal is to explain policy dynamics from the early 1980s to present, focusing not only on party preferences, but also agenda capacity and the content of parliamentary activities. To do this, we explore changes in agenda attention over time, whether agenda content is determined by which party is in government, how perceptions about economic and political conditions influence agenda capacity and composition, and to what extent changing rules in the Congress of Deputies affects the prioritization of issues.

Increasing confrontation and polarization has not generated an intensification of parliamentary activities. Overall, the analysis of issue attention in the Spanish Parliament shows that political parties have a minor interest in increasing the importance of non-parliamentary activities as would be expected following the logic of issue competition (Green-Pedersen, 2007; Green-Pedersen and Mortensen, 2010; Mair et al., 2004). In this view, political parties would use parliamentary activities as a means to reinforce issue ownership, emphasizing their concern by those issues they own. The basic assumption is that parties are rational actors that can identify the issues that can maximize their electoral rewards. Parties would focus their attention in the issues they own, forcing the government to address issues it might prefer to ignore in the parliamentary debate.

However, in this chapter we explain the monopolization of the political debate concerning some issues is a conditional phenomenon, occurring only when a party is not governing. Overall, when asking oral questions the PP tends to highlight economic and governmental issues, while the PSOE significantly asks more questions about welfare and civil rights issues. This pattern of issue specialization in the parliamentary arena corresponds with the general opinion of Spanish citizens about which party is more capable to deal with which set of issues.[1] Still, our results indicate that for many issues, the strategy of issue ownership is replaced by one of embarrassing the government, forcing it to discuss policy failures or politically unpopular issues (Green-Pedersen and Mortensen, 2010). Both the PP and PSOE tend to highlight different issues depending on whether they are governing or not, and this is not necessarily linked with the executive's policy decisions but unexpected and focusing events like the Iraq war and economic conditions.

Governmental status also explains variations in agenda fragmentation and the use of non-ordinary questions for the case of statewide parties. Both the PP and PSOE tend to concentrate their attention on fewer issues and increase the number of urgent questions when they are in the opposition. This is a means to force the government to take a position about issues it might prefer to ignore. This is especially the case after changing the rules of oversight activities in the Spanish Parliament in 2008. This institutional reform transformed the functioning of non-parliamentary activities in Spain from one in which the party in government and the main opposition party asked the same number of questions, to a new situation in which the incentives of the governing party to actively participate in the question period were reduced to the minimum. As a result, by 2014 oversight activities were increasingly monopolized by opposition parties and occupied a much smaller share of parliamentary activities than eight years before.

To explain all this, the chapter is organized as follows. The first two sections explain how formal rules and governmental status affect agenda capacity, agenda diversity, and the type of questions. First, we provide detailed information about the two major reforms about oral questions in the Spanish Parliament, and from here we explain how they affect the number and type of questions asked by each parliamentary group and the degree of agenda fragmentation across time. Also, we show that governmental status is a key variable to explain the type of questions and agenda fragmentation, but not the number of questions before 2008. The third part explains variations in issue attention across time and explains whether this depends on governmental status.

Formal rules governing oral questions: agenda capacity

Agenda capacity affects the ability of MPs to push a new issue onto the formal agenda (Jones and Baumgartner, 2005). The fact that an MP introduces an oral question about violence against women may be related to a new case of a woman murdered by her husband, to increasing media attention on these issues due to citizen mobilization, and/or party preferences. Or the MP may long have wanted to ask the question but other more pressing or urgent political matters forced them to use their limited allotment of questions to ask a different one. MPs have to prioritize among issues because agenda capacity is limited. They must decide whether to emphasize violence against women vis-à-vis a question about immigration, education, or energy policy. In doing so, they take into account the importance of this issue for their constituencies, the position of the party in government, and the number of questions MPs can introduce per session.

Formal rules and procedures constrain agenda capacity. As Table 4.1 illustrates, oral questions are presented at a fixed question time on a regular basis (generally on Wednesdays). MPs hand in their questions

Table 4.1 General features of plenary oral questions in the Congress of Deputies

<table>
<tr><th>Period</th><th>1982–1996</th><th>1996–2008</th><th>2008–2015</th></tr>
<tr><td>OQ are registered and asked by</td><td colspan="3">Individual MPs, although
The party speaker has a predominant role</td></tr>
<tr><td>OQ Allotment</td><td>Not defined</td><td>Maximum of 24</td><td>Maximum of 25</td></tr>
<tr><td>Type of oral questions:</td><td></td><td></td><td></td></tr>
<tr><td>(1) Ordinary</td><td colspan="3">Questions about political events and/or decisions registered by Thursday</td></tr>
<tr><td>(2) Non-ordinary</td><td>About political events and/or decisions happening during the weekend registered Friday to Monday 12:00 noon (from 1986).</td><td colspan="2">(a) Council of Minister questions defined as those registered on Friday before 8:00 p.m., dealing with decisions of the Council of Ministers and (b) urgent questions defined as those registered after Friday and before Monday 12:01 a.m.</td></tr>
<tr><td></td><td colspan="3">From 1996, to register and ask non-ordinary questions always imply the substitution of an ordinary question.</td></tr>
<tr><td>General requirements to register an OQ</td><td colspan="3">They may contain only the bare and narrow formulation of a single issue, questioning about an event, a situation or information in which the Spanish executive is involved.</td></tr>
<tr><td rowspan="3">Specific requirements to register an oral question</td><td colspan="3">Ordinary questions: parliamentary board supervision. OQ must fulfill all formal general requirements in terms of length and content.</td></tr>
<tr><td rowspan="2">Non-ordinary OQ: the substitution of an ordinary by a non-ordinary OQ requires the agreement of the Speakers Board and the executive.</td><td colspan="2">Council of Minister OQ: parliamentary board supervision. OQ must fulfill all formal requirements in terms of length and content.</td></tr>
<tr><td colspan="2">Urgent OQ: the substitution of an ordinary question by an urgent question requires the agreement of the Speakers Board and the agreement of the executive.</td></tr>
</table>

Party allotment	Not-fixed allotment. In practice, the speakers and parliamentary board define the allotment of questions per party taking into account the number of seats.	Not-fixed allotment. Oral questions are distributed as follows: (a) all parliamentary groups can ask at least one question; and (2) one question every 10 seats or fraction.	Fixed allotment by parliamentary group. From 2008 to 2011 the distribution is as follows: PSOE (10); PP (10); Catalan (CiU): 1; Basques (EAJ-PNV): 1; IU-ICV-EUiA, CHA (izquierda plural): 1; UPyD (1), Mixto: 1.
Question time	Once a week. Generally on Wednesday from 1994. About five minutes per question: first an MP asks a question, then a governmental actor (Prime Minister, a minister or a state secretary) responds. MPs have the right to respond. After that, members of government may say the last word.		
Exceptions	The executive may ask for one week extension to answer an oral question. Exception: the parliamentary board may cancel the questioning session, or limit the number of questions.		

Source: Reglamento del Congreso (www.congreso.es) and *Boletín Oficial de las Cortes Generales* (BOCG). Resolución de la Presidencia del Congreso de los Diputados sobre el desarrollo del artículo 188 del reglamento del Congreso, de 12 de enero de 1983. Resolución de la Presidencia del Congreso de los Diputados, de 18 de Junio de 1996, sobre el Desarrollo del artículo 188 del Reglamento, relativo a pregunta para respuesta oral en Pleno. Resolución de la Presidencia del Congreso de los Diputados, de 10 de Junio de 2008, sobre desarrollo del artículo 188 del Reglamento, relativo a preguntas para respuesta oral en Pleno (modificada por Resolución de la Presidencia, de 28 de Diciembre de 2011).

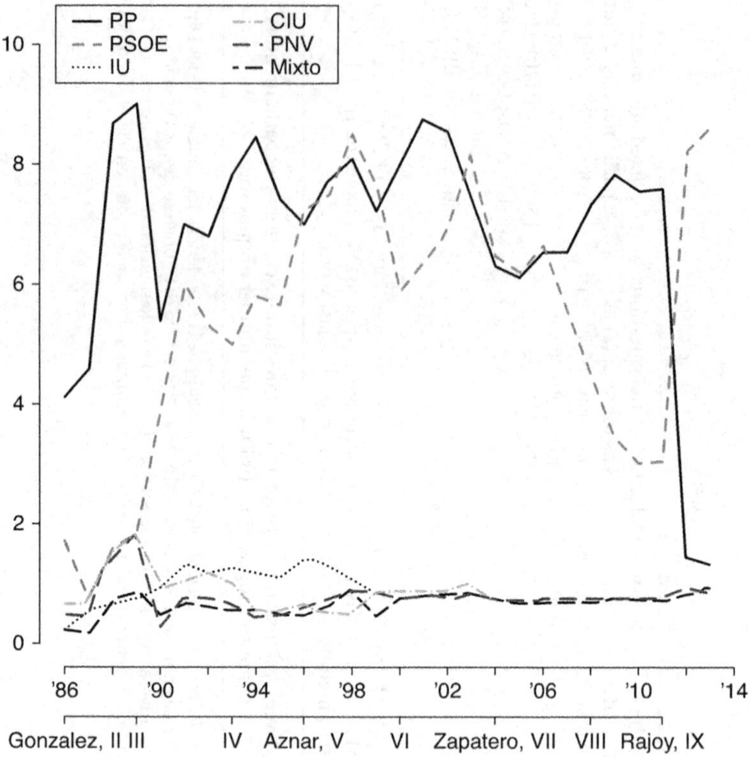

Figure 4.1 Average number of questions asked per session, by party

Note: For each political party, the figure shows the average number of questions posed by its deputies during weekly question time.

Source: Spanish Policy Agendas Databases (www.ub.edu/spanishpolicyagendas).

in advance so that the executive has time to prepare a response. When question time comes (the following Wednesday), the MP asks the question (orally) and a member of the executive branch (the Prime Minister, a Minister, or a Secretary of State) responds (orally). Individual MPs, not parliamentary groups, have the authority to ask parliamentary questions. However, in practice, the strong party discipline that characterizes the Spanish political system means that the parties control the process.

The rules governing question time have been significantly changed three times, as Table 4.1 makes clear. Initially, from 1982 to 1996, there was no fixed number of questions per session and/or per party. The speakers and Parliament board decided which Parliamentary Group (PG) could introduce questions; they acted as gatekeepers of each session. Since 1996, greater structure has been imposed.[2] The number of questions was set at 24 questions per session (changed in 2011 to 25), and every PG has the right to ask at least one question; the remaining questions are distributed according to the number of seats – one oral question per every ten seats or fraction thereof. In 2008, another amendment to the rules further transformed the distribution of oral questions among the party groups: 10 oral questions for the two large parliamentary groups, the PSOE and PP; and one question for any other parliamentary group in the Chamber of Deputies. In this last reform, the rules (Art. 6.3) explicitly prohibit PGs from asking more questions than their fixed allotment, even when other PGs do not use their fixed allotment.

These changes in formal rules have a direct impact on the goals and strategies MPs pursue in the parliamentary arena. First, the overall number of questions (registered and asked) became more predictable over time, especially after 1996 when the rules of Parliament first defined a fixed number of questions per group. Second, the number of question periods scheduled but cancelled was reduced to almost zero. From 1983 to 1996, about 170 sessions were scheduled but did not occur (22 percent of the total sessions scheduled), while from 2008 to 2013 this only occurs sporadically. The reasons to cancel an oral question session are loosely defined. In practical terms, this means it is the parliamentary board, in cooperation with the government, which decides whether a session is going to be held or not. One of the consequences is a high percentage of questions that were registered but never asked (about 40 percent) in the early 1980s.

Third, the mean number of questions per session increased from an average of 15 questions per session during the period 1983–1996 to 18 questions during the period 1996–2008, to decline again to an average of 14 questions per session for the period 2008 to 2013 (Figure 4.2). Fourth, from 1996 to 2008, the mean number of questions asked by the governing and main opposition party are practically the same, while from 2008, differences in the number of questions asked by the two main parties increased dramatically. Finally, changing formal rules had no significant impact on the average number of questions asked by small and regional parties. Agenda capacity of parties like IU, CiU, or the PNV is stable and limited to about one question per session, which as we explain in the following sections, directly implies a much more concentrated issue agenda than large statewide parties like the PP and PSOE.

Overall, fluctuations in the number of oral questions are especially relevant in the case of the governing party, but not opposition parties. After

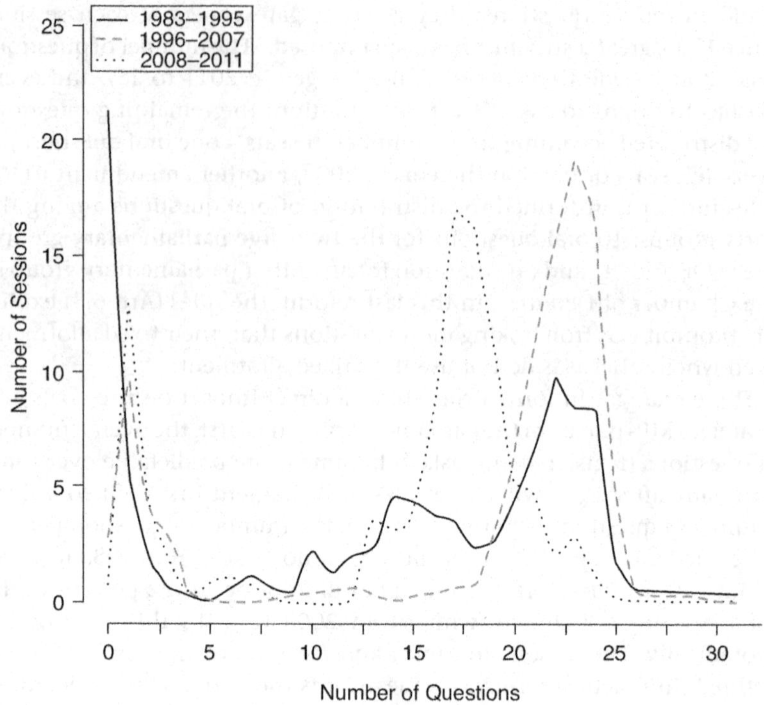

Figure 4.2 Number of questions per session

Note: For each of the legislatures listed (1983–1995, 1996–2007, and 2008–2011), the lines indicate how many sessions saw a given number of questions posed to the government. Typically, 15 to 25 questions are posed, though there are cases with no questions, or just a few.

Source: Spanish Policy Agendas Databases (www.ub.edu/spanishpolicyagendas).

taking control in 1996, the PP asked the same number of questions on average as when it was in the opposition. In contrast, after winning the majority in 2011, the PP only asked one question on average per session. This was eight times less than the number of questions asked while José Maria Aznar was governing (1996–2004), and only 10 percent of its full allotment at that time. The same occurs with the PSOE. During the almost 14 years of government of Felipe González, MPs of the PSOE only asked 4.6 questions (average) per session; almost half the oral questions asked during the first legislature of Rodríguez Zapatero (2004–2008). Once rules changed in 2008, PSOE oral questions declined to 3.5 per session.

For decades, the governing party has used most its allotment of questions (10 out of 24 (later 25), or about 40 percent of the total) to highlight the benefits of governmental decisions, more than controlling whether the executive was following its mandate. Actually, this is what one might expect to happen in a political system characterized by blocked lists and strong party discipline in which MPs of the governing party have little autonomy to criticize any of the policy decisions taken by the executive, even in the case of moral issues. From 2008, the incentives of MPs of the governing party to ask questions decline and oversight sessions become increasingly a political venue in which opposition parties take the lead. As we explain in the next section, this is very much related to changing rules in 2008, which explicitly prohibited that parliamentary groups ask more questions than their fixed allotment.

Explaining variation

The description so far calls into question existing explanations about the growing importance of non-legislative activities in parliamentary democracies (Green-Pedersen, 2010; Franklin and Norton, 1993; Wiberg, 1995; Russo and Wiberg, 2010; Sánchez de Dios and Wiberg, 2011). In particular Wiberg (1995) argues that the increasing importance of parliamentary questioning is linked to the growing complexity of advanced democracies and the expansion of the public sector. The more the scope of state intervention the more the time MPs have to devote to oversight sessions in order to ask the executive about what it is doing about a vast variety of issues. Other authors, including Green-Pedersen (2010) stress that parliamentary questions increase as a result of a rational strategy of political parties designed to reinforce issue ownership. MPs would increasingly engage in parliamentary questions as a means to emphasize their concern by those issues they "own," and which are thus most rewarding in electoral terms. Both arguments stress that the number of oral questions would increase across time, either as a result of state intervention or party strategies.

To test these arguments we developed a simple OLS regression model in which the dependent variable is the number of oral questions asked in each month from 1986[3] to 2013, and the three independent variables are: (1) time (measured by year, with positive coefficients indicating oral questions increase across time); (2) rules (measured as a dummy variable scored 0 before 2008 and 1 in 2008 and onwards, with negative coefficients indicating oral questions decline after 2008); and (3) the public mood (see Chapter 2 for a detailed description). With this variable we measure to what extent changing perceptions on the economic situation affect the number of questions MPs are asking across time.

We expect that the more negative the public mood, the more MPs will use oral questions as a means to attack the governing party, forcing the executive to take a position about issues it would prefer to avoid.

Results are summarized in Table 4.2. Contrary to existing explanations, our results illustrate parliamentary questions are declining in Spain for the last two legislatures, and this is very much related to changes in the formal rules governing parliamentary questions from 2008 and negative public mood. First, coefficients for the variable "rules" are negative, and significant (at .01), which means that the overall number of questions declines by 13.2 after 2008. Second, as expected, the number of questions decline as there is a growing percentage of citizens that perceive the economic and political conditions as positive, although results here are not significant. Finally, coefficients for the variable "time" are also negative and significant at .005, which contrary to existing analysis indicates oral questions tend to decline across time, even controlling for rules changes and the level of public concern about the state of the economy.

In order to go further in the analysis we test whether political parties follow a different strategy depending on whether they are governing or not. Here, we run two separate equations for each political party, in which the dependent variable is the number of questions asked by the PSOE and the PP, respectively, from 1986 to 2013. We limit the analysis to large statewide parties, mainly because in contrast to any other party, they have more than one question per session (a maximum allotment of ten questions); and also, because both parties have been governing at some point during that period. In both cases we introduce three additional independent variables to the previous model: (4) governmental status, this is the institutional position of the party in the governing-opposition game (measured with a dummy variable: 1 if governing, 0 if in the opposition), (5) an interaction term designed to measure whether changing rules in 2008 affect political parties differently depending on whether they are governing or not, and (6) an interaction term designed to measure whether changes in the public mood affect a political party differently depending on whether it is governing or not (see Brambor et al., 2006 for methodological discussion). All calculations are at a monthly level.

Results are summarized in Table 4.2. First, as in the previous model, the number of oral questions asked by the PP and the PSOE decline after 2008, and also when the public mood is perceived as positive. Second, the impact of changing rules is more relevant to explain the PSOE oversight activities (oral questions decline by 16.45, significant at .003), than

Table 4.2 Explaining variation in the number of oral questions

	B	Sig.	R^2
PP			.270
Rules	–4.811	.179	
Mood	–14.774	.021	
Time	–.009	.949	
PP governing	–11.802	.170	
Mood x PP governing	19.932	.139	
Rules x PP governing	–15.494	.029	
PSOE			.267
Rules	–16.449	.003	
Mood	–10.121	.294	
Time	.724	.000	
PSOE governing	–6.326	.365	
Mood x PSOE governing	.755	.945	
Rules x PSOE governing	–.716	.901	
All Parties			.069
Rules	–27.514	.001	
Mood	–5.790	.617	
Time	.649	.035	

Note: Entries are the result of an OLS regression covering the period of 1986 through 2013. The dependent variable is the number of oral questions introduced (PP, PSOE, and all parliamentary groups). The independent variables are rules (0 from 1986 to 2008, and 1 after 2008), variations in the public mood (positive values indicating citizens' positive perception of the economic and political situation), time (years, from 1986 to 2013), government-opposition status (1 if the party in question is governing), and two interaction terms controlling for the impact of governing status on the mood and rules variables.

the PP's (oral questions decline by 4.8, but not significantly). In contrast, the public mood seem to be especially relevant to explain the evolution of the oral questions asked by the PP (oral questions decline by 14.7, significant at .02 when the economy is perceived as positive), and only to a lesser extent for the PSOE (oral questions decline by 10.12 but not significant). Third, in both cases, the number of oral questions tends to decline when the party is governing.

Still, here what is especially interesting are the results of the coefficients for the interaction term "governing-public mood." These coefficients indicate that in both cases, the impact of the public mood varies depending on whether they are governing or not. When the PP is governing and perceptions about economic and political conditions are negative, the number of PP oral questions decreases by 19.93 (significant at .1). By contrast, when the PP is not governing and the public mood declines, the number of questions increase. In the case of the PSOE, the argument is the same but coefficients are not statistically significant. Finally, in both cases the number of oral questions decline after 2008, but even more strongly when the party is governing.

In sum, results indicate that changing rules and public perceptions about the economy and political conditions affect the strategies of political parties when performing their tasks in the oversight sessions, but they do so with different intensity depending on their position in the government-opposition game. During the two legislatures of Rodríguez Zapatero, the PP kept asking a similar number of questions as before, until it won the elections in 2011. By contrast, the number of questions asked by the PSOE decreased after 2004, only to increase again after Mariano Rajoy won elections in 2011, once it occupied the role of main opposition party. Also, these results emphasize the importance of economic and political perceptions on the government-opposition game. The better the public mood, the less the number of questions asked by political parties, but only when they are in the opposition. In the next section we explain whether changing rules affect parties' strategies across all types of questions.

Diverse foci of attention for routine and urgent parliamentary questions

From 1986 through 1996, MPs could ask only two types of oral questions: routine questions, which are registered in the parliamentary agenda by Thursday; and non-routine questions, which refer to those registered after Thursday. All questions are then posed orally the following Wednesday, as Table 4.1 made clear. Beginning in 1996, non-routine questions are further divided into (a) Council of Ministers questions, which must be registered by Friday at 8:00 p.m.; and (b) urgent questions, related to issues happening after Thursday (and not directly related to the Council of Ministers), which must be registered before Monday at 12:00 noon. Figure 4.3 shows how fluctuations in the trends of parliamentary questions differ for non-routine and routine questions.

Different question types serve different purposes such as obtaining information from the government, holding the executive accountable, raising a constituency matter, compelling the Prime Minister to act on a certain matter, forcing the government position on a highly controversial matter, or providing the government an opportunity to publicize its opinion about an issue (Wiberg, 1995: 181). Members of the governing party use ordinary questions to highlight policy achievements and

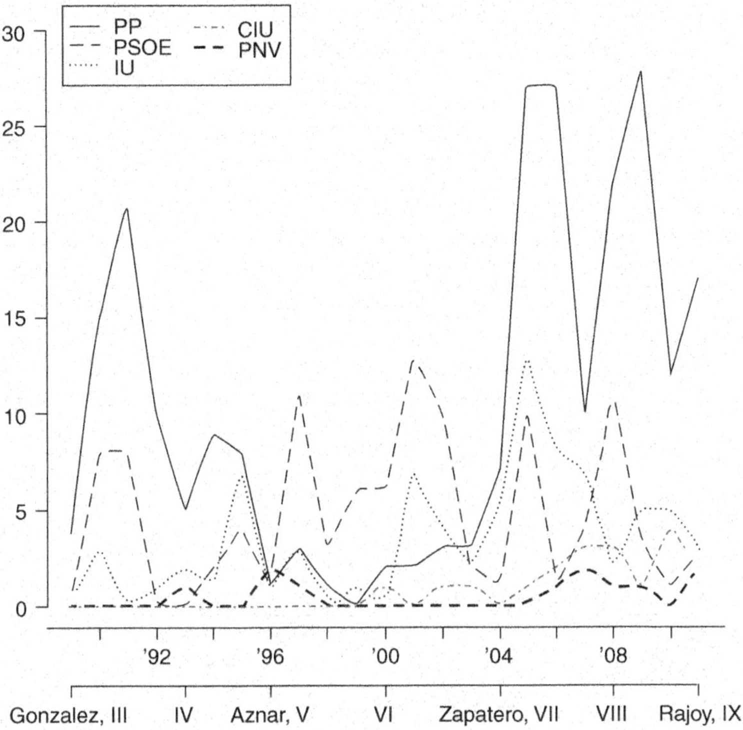

Figure 4.3 Number of non-ordinary questions, by year and by party

Note: For each party, the figure indicates the number of Urgent or Council of Ministers questions posed in each year.

Source: Spanish Policy Agendas Databases (www.ub.edu/spanishpolicyagendas).

successful reforms. Party discipline guarantees most oral questions introduced by the governing party MPs are arranged questions that respond to a planned strategy designed to minimize the political costs associated with certain governmental decisions. In contrast, urgent questions are the perfect tool for opposition parties seeking to highlight problematic issues, especially those capturing media attention. Finally, Council of Ministers questions were designed in the 1996 reform as a means to promote weekly discussion of the decisions taken in Cabinet meetings each week. Unlike the rules governing question time before 1996, this reform allowed MPs to substitute a routine question with a Council of Ministers question, without approval by the Speaker's board.

Overall both the PSOE and PP ask more non-ordinary questions, including urgent and Council of Ministers questions, when they are in the opposition (Table 4.3). In the case of the PP, the average number of non-ordinary questions increases from 2.11 to 10.65; and for the PSOE, increases from 2.55 to 5.44 when they move from government to opposition (mean differences significant for all cases). In the case of small and regional parties, results illustrate oral questions most of the time follow ordinary procedures. These results clearly emphasize the function of non-ordinary questions as an instrument of the main opposition party to highlight controversial issues with an important potential to erode the party in government.

The next question is whether variations in the number of non-ordinary questions are also explained by changing rules. To test this, we run another OLS regression model in which the dependent variable is the number of non-ordinary questions asked by party each month, and the independent variables are: (1) rules, measured by a dummy variable in which 0 is the period before 1995 and 1 is the period after 1995 (note that

Table 4.3 Oral questions posed by party group, by opposition status

Mean number of oral questions posed by year	Opposition	Government	Mean difference	Sig. (t-test)
PP total	191.3	160.9	30.4	.167
PP ordinary	180.65	186	–5.3	.777
PP non-ordinary	10.65	2.11	8.5	.000
PSOE total	186.64	100	86.6	.000
PSOE ordinary	170.22	98.25	72	.004
PSOE non-ordinary	5.44	2.55	2.9	.083
IU total	23.09	20.25	2.8	.497
IU ordinary	21.23	13.25	8	.05
IU non-ordinary	1.86	7	–5.1	.000
CiU total	29.3	26.22	3.1	.516
CiU ordinary	21.85	19.11	2.7	.343
CiU non-ordinary	.2	.22	.0	.913
PNV total	18.33	18.4	–.1	.984
PNV ordinary	15.75	14.6	1.2	.73
PNV non-ordinary	.17	0	.2	.452

Note: The table shows the mean number of oral questions of different types and compares, for each party, the number posed while they are in opposition versus in government (or supporting a government even while not actively participating in it). High values in "mean difference" mean that the party asks fewer questions while in government than while in opposition. Data are annual from 1986 through 2011.

Source: Spanish Policy Agendas Databases (www.ub.edu/spanishpolicyagendas).

here the major transformation occurs with the reform of parliamentary rules in June 1996); (2) governmental status, the institutional position of the political party in the governing-opposition game (1 = governing or giving support to the government, and 0 = non-governing); and (3) an interaction terms designed to measure whether changing rules in 1996 affect political parties differently depending on whether they are governing or not. We run the same regression for the two large political parties (the PP and PSOE) and the three other parties that have a parliamentary group for the whole period under analysis (1986 to 2011).

Results are summarized in Table 4.4. First, in all cases, once the number of veto points was reduced in 1996, political parties ask more non-ordinary questions. Before 1996, to register a non-ordinary question – either urgent or Council of Ministers question – required the substitution of an already registered ordinary question, and the unanimous agreement of

Table 4.4 Explaining the use of non-ordinary questions

	B	Sig.	R^2
PP			**.109**
Rules x Governmental status	1.233	.18	
Rules	.858	.015	
PP governing	–2.283	.000	
PSOE			**.014**
Rules x Governmental status	.708	.216	
Rules	.255	.241	
PSOE governing	–.063	.815	
IU			**.158**
Rules x Governmental status	–.384	.071	
Rules	.169	.057	
IU support	.642	.000	
CiU			**.034**
Rules x Governmental status	–.131	.343	
Rules	.112	.041	
CiU support	–.140	.009	
PNV			**.026**
Rules x Governmental status	–.191	.015	
Rules	.052	.093	
PNV support	.059	.172	

Note: The dependent variable is the number of non-ordinary oral questions introduced by each parliamentary group. The independent variables are variation in the public mood (positive values indicating citizens' positive perception of the economic and political situation), rules (scored 0 from 1986 to 2008, and 1 after 2008), and the government-opposition status of the party in question. OLS coefficients are shown. The model also included a constant term, not shown. Analysis is annual from 1986 through 2011.

parliamentary groups (speakers board). This meant the governing party could temporarily veto the question and thereby lessen the level of politicization of a specific issue. Of course, this is only a temporary measure, as the same MP could inquire about the matter as a routine rather than an urgent question in the following week. Still, timing matters in politics so this reform was significant in enhancing the power of the opposition. After 1996, unanimity is only required in the case of urgent questions introduced after Friday. This means MPs can freely decide whether to substitute an ordinary question with a Council of Ministers question. Actually this modification has a positive and statistically significant effect on all parties except the PSOE (for which results are positive but not significant). The reduction of veto points is especially relevant for small and regional parties, which only rarely introduce a single urgent question from 1986 to 1996.

Second, results indicate party status, whether governing or in the opposition, are also relevant to explain variations in the number of urgent questions. Coefficients are negative for the case of the two large parties and CiU, but not significant in the case of the PSOE. This means that when the PP is in the opposition, the number of non-ordinary questions increase by 2.2 (significant at .000), and when CiU is giving support to the party in government, non-ordinary questions decrease by .140 (significant at .05). In short, as expected, results thus far indicate that (1) parliamentary groups strategically select the type of questions, renouncing non-routine questions when they are governing; and (2) political parties tend to use more and more urgent questions as the number of veto points decreases after 1996. Our next question is whether political parties are also strategically selecting the type of issues they talk about in the parliamentary arena.

Issue competition and agenda fragmentation

Mandated limits on the agenda capacity of the oral question sessions in the Spanish Parliament force MPs to be selective in the issues they discuss and champion. Small and regional parties must be pickier in their parliamentary activities than large statewide parties, because their agenda capacity is smaller. Because CiU, PNV, and CC are only allotted one question per session, they have strong incentives to concentrate on only those issues their constituencies identify as most important such as transfers of issue jurisdiction, education, and transportation in the case of the regional parties, or labor and defense issues in the case of the Far Left. Other issues may occupy a secondary position, especially when regional and small parties share the same policy position as the governing party. For example, regional parties like ERC or BNG shared common policy positions on moral issues with the PSOE government of Rodríguez Zapatero. Both were important policy entrepreneurs adopting an active role for policy change in the parliamentary arena prior 2004, but once the PSOE won elections, they did not introduce a single question about same-sex marriage, abortion, or stem-cell research.

While small and regional parties have to horde their attention, national parties may diversify their agenda attention due to their larger agenda capacity – a maximum of ten questions per week. That is, their agenda would be more diversified basically because their allotment of questions is larger. To measure agenda fragmentation we computed Shannon's entropy scores from 1986 to 2011 for the two large statewide parties, and the three parties with a permanent parliamentary group from 1986 to 2011 (Boydstun et al., 2014). Figure 4.4 clearly illustrates fragmentation

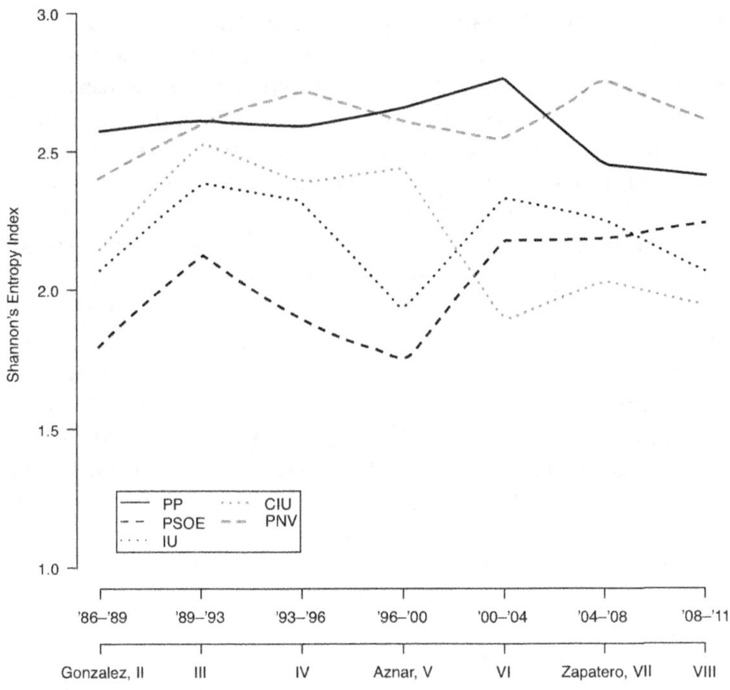

Figure 4.4 Spread of attention in parliamentary question time, by party

Note: For each party, the figure shows Shannon's entropy index, a measure of spread of attention. Higher values indicate greater spread of attention; lower values, attention concentrated on a small number of topics.

Source: Spanish Policy Agendas Databases (www.ub.edu/spanishpolicyagendas).

is larger for statewide parties than for small and regional parties, with important variations across time, which seems to be connected with party strategy more often than changing rules. While in the opposition, one may expect that statewide parties would be willing to concentrate issue attention on those problems that are more controversial, such as the rescue of bankrupt financial institutions, or political decentralization, leaving aside those issues for which there is an agreement among political parties, like terrorism from 2000 to 2004 (see Chapter 3).

In contrast, MPs of the governing party will diversify their agenda across issues, asking members of the Spanish government about policy reforms and achievements of each executive area, from health to social policy, and EU matters especially when there is no conflict involved. Additionally, we may expect issue diversity will be especially high before 2008 when the rules governing oral question periods generate important incentives for governing parties to use their full allotment of questions. Finally, in the case of small and regional parties, agenda fragmentation would not vary much across time, mainly because these parties only have one question per session without much variation across time.

To test this hypothesis we ran an OLS regression in which the dependent variable is agenda fragmentation (Shannon's entropy scores by year and by party), and the independent variables are: (1) governmental status and (2) rules,[4] both defined as in the previous models. The results summarized in Table 4.5 clearly give support to the assumption that parties have a more fragmented agenda when they are governing than while in opposition – entropy scores increase by .160 and .103 when the PP and PSOE, respectively, are governing (significant at .001 and .01). Also, results illustrate that fragmentation decreases after changing rules in 2008 for both parties (.084 in the case of the PSOE and .122 in the case of the PP, significant at .06 and .09). In the case of small and regional parties, fragmentation tends to decline when they are giving support to the governing party and from 2008, but results are not significant for any of the cases.

In short, results so far indicate that statewide parties, as rational actors, decide whether or not to diversify issue attention according to both their issue preferences and the potential benefits associated with a given strategy. When they are governing, oral questions tend to be more diversified across issues, while the opposite occurs when parties are in the opposition. Unexpected events and policy failures concentrate most of the attention of opposition parties as a means to erode the governing party, forcing it to take policy positions about highly controversial issues like the Prestige oil spill, the Iraq war, or rising unemployment. Next,

Table 4.5 Explaining agenda fragmentation

	B	Sig.	R^2
PP			**.439**
Rules	–.14	.01	
PP governing	.138	.000	
PSOE			**.087**
Rules	.214	.167	
PSOE governing	.084	.600	
IU			**.002**
Rules	.01	.98	
IU giving support	.092	.83	
CiU			**.047**
Rules	–.15	.31	
CiU giving support	–.08	.48	
PNV			**.066**
Rules	.424	.19	
PNV giving support	.006	.98	

Note: The dependent variable is agenda fragmentation (Shannon's entropy score). The independent variables are rules (scored 0 from 1986 to 2008, and 1 after 2008) and the government-opposition status of the party in question. OLS coefficients are shown. The model also included a constant term, not shown. Analysis is annual from 1986 through 2011.

we question whether political parties increasingly engage in a calculated issue-selection process, as a new form of party competition. Following the issue-competition literature, we analyse how issue attention varies across time and political parties in Parliament.

Increasing issue competition

What issues are parties talking about when asking oral questions? Table 4.6 summarizes mean differences across the 19 CAP issues for the case of the PP and PSOE. It shows that parties show different levels of concern about issues over time. Overall, the PP emphasizes issues related to the functioning of the economy (macroeconomics), and governmental issues, especially political decentralization. In contrast, the PSOE asks significantly more questions about rights issues, health, education, the environment, and housing. For all other issues, differences are not statistically significant.

This pattern of issue specialization corresponds with the general opinion of Spanish citizens about which party is more capable to deal with which set of issues. According with the poll conducted by the

Table 4.6 Oral questions introduced by the PP and the PSOE across topics

	Mean percentage of oral questions introduced		Difference	
Topic	**PP**	**PSOE**	**PP-PSOE**	**Sig. (t-test)**
Economy	6.583	2.710	3.873	.000
Rights	1.896	3.150	−1.254	.006
Health	4.269	5.621	−1.352	.032
Agriculture	4.512	4.499	.013	.981
Labor	2.427	2.661	−.234	.563
Education	5.318	6.563	−1.245	.050
Environment	1.792	2.768	−.976	.017
Energy	2.089	1.831	.257	.429
Transportation	6.600	6.174	.426	.555
Crime & justice	7.628	7.000	.628	.388
Social policy	1.680	1.821	−.142	.673
Housing	1.108	2.541	−1.434	.000
Commerce	3.360	3.124	.236	.644
Defense	3.432	3.411	.021	.974
Research & development	2.628	3.377	−.749	.082
Foreign trade	.484	.505	−.022	.883
International affairs	4.224	4.463	−.239	.660
Government	9.260	5.583	3.677	.000
Public lands	1.814	1.971	−.157	.629

Note: Data refer to the period of 1983 through 2011.

Source: Spanish Policy Agendas Databases (www.ub.edu/spanishpolicyagendas).

Centro de Investigaciones Sociológicas[5] in 2009, Spanish citizens considered the PP more capable of managing economic issues – this is the case for 43 percent of PSOE voters and 76.2 percent of PP voters. The opposite occurs with social policy and the environment; in both cases the majority of the Spanish population considers the PSOE more capable of handling these problems. This is even the case for PP voters (only 49 percent of PP voters consider the PP will do a better job than the PSOE in relation to social policy, and only 43 percent for the case of environment).

Next, we turn to the question of whether there are important variations in the pattern of issue prioritization across time. Political parties, when performing their tasks in parliamentary oversight sessions, should pay attention to those issues the executive is talking about or taking decisions about. That is, one may expect the PP would ask oral questions about rights issues like abortion or same-sex marriage when Rodríguez Zapatero introduced major regulatory reforms, or that the PSOE would increasingly ask questions about macroeconomics, the privatization of public companies, the liberalization of telecommunication services, or defense issues after the PP wins elections. Both rights and the economy are issues traditionally associated with the PSOE and PP, respectively. However, neither the PP nor the PSOE can avoid paying attention to these issues when they become one of the centerpieces of the governmental agenda and their party controls the government. The same argument applies for the case of focusing events. Political parties have to show concern about issues like the Iraq war, economic recession or environmental catastrophes independently of whether they "own" these issues or not.

Also, according to the issue-ownership approach, we should expect political parties to focus their attention on the issues they "own," leaving aside those that most citizens do not associate with them. This means issue attention would follow a quite stable pattern for a long period of time. The PP will focus its attention especially on economic issues, the PSOE on social issues and the environment, regional parties on political decentralization, and the far-left on social issues, defense, and labor – independently of whether they are governing or giving support to the party in government, the evolution of the economy, and/or agenda capacity. To assess this argument, we compute mean differences of issue attention when the PP or the PSOE are governing or in the opposition. Results are summarized in Table 4.7.

One of the main conclusions is that both political parties change their priorities depending on whether they are governing or in the opposition. First, in both cases, MPs pay significantly more attention to welfare-related issues (mainly health and social policy) and infrastructure (transportation and water policy) when they are governing. For some issues, such as transportation, both parties may obtain electoral rewards for giving publicity about governmental decisions in the parliamentary discussion. The construction of a new high-speed train line connecting Madrid with any other territory in Spain may have important and positive electoral rewards for the governing party in that particular territory, independent of the economic costs associated with this decision, which generally are quite opaque for most citizens (Bel, 2011). In fact, starting in the early 1990s, various Spanish governments have devoted a large amount of public resources to the construction of new airports – there are more than 50, all of them operating with budget deficits with the exception of Palma de Mallorca and Gran Canaria – and to enlarge

Table 4.7 Change in attention, by government-opposition status

Topic	PP	Sig. (t-test)	PSOE	Sig. (t-test)
Economy	.129	.906	4.232	.000
Rights	.773	.139	–.206	.811
Health	–3.290	.000	–2.187	.132
Agriculture	–1.083	.183	–.480	.605
Labor	–.173	.752	1.965	.009
Education	–.949	.261	–.659	.572
Environment	–1.680	.000	.521	.528
Energy	.873	.065	1.205	.032
Transportation	–1.134	.229	–3.778	.005
Crime & justice	4.698	.000	3.042	.010
Social policy	–.678	.080	–.977	.098
Housing	–1.034	.005	.526	.497
Commerce	.048	.936	–.658	.489
Education	–.766	.353	.426	.730
Research & development	–2.838	.000	.770	.524
Foreign trade	–.034	.896	–.656	.105
International affairs	–1.138	.203	–1.290	.143
Government	9.157	.000	6.028	.000
Public lands	–1.772	.000	–.915	.090

Note: Entries show the difference in the percent of questions posed by a given party on the topic shown when the party is in the opposition compared to when they are governing. Data cover the period of 1983 through 2011.

Source: Spanish Policy Agendas Databases (www.ub.edu/spanishpolicyagendas)

the network of high-speed trains (see Bel, 2011). In 2014, Spain has the second-longest network of high-speed trains after China, in spite of the fact that most lines operate in the red. Both the PP and PSOE have taken electoral advantage of the construction of these enormous infrastructure projects, which by definition have highly diffused costs and very concentrated benefits.

In other cases, like social policy, both the PP and PSOE ask more questions when they are governing as a means to give publicity to policy decisions that generate no controversy (e.g., the implementation of family policy by the PP or the *Ley de Dependencia* by the PSOE). Deputies of the PP and the PSOE introduce oral questions as a means to emphasize those governmental decisions which have the support of most Spanish citizens and directly affect some of the most vulnerable constituencies, like disabled people or families living in poor economic conditions.

Second, there are some issues for which the PP and PSOE follow a different strategy depending on whether they are governing or in the opposition. In the case of defense, the environment, labor, research and development, and macroeconomics, the PP asks more questions when governing and the PSOE asks more questions while in the opposition. The opposite occurs in the case of rights and commerce and banking; in both cases, the PSOE asks more questions when governing. MPs of the PSOE ask questions to the PP governments about the entry into the war of Iraq, the environmental catastrophe of the Prestige oil spill, the liberalization and re-regulation of telecommunications in the late 1990s, and/or the entry of Spain into the Economic and Monetary Union (EMU). In contrast, while in the opposition, the PP asks more questions about commerce and banking, such as the banking reform developed by Rodríguez Zapatero in the late 2000s; and civil rights issues, mainly regulatory reforms surrounding moral issues.

Finally, in the case of crime and justice, governmental issues, and energy, both parties significantly ask more questions when they are not governing. In the case of crime and justice, mean differences are especially intense in the case of the PP and this is linked to terrorism. The PP is especially active on terrorism when it is not governing – oral questions of the PP about crime and justice increase by 4.2 when it is not governing – especially from the mid-2000s, when the number of ETA victims declined to almost zero (see Figure 3.4 in Chapter 3). Actually the PSOE did not ask a single question about terrorism from 2000 to 2004 as a result of the agreement between the PP and PSOE to fight against terrorism (see Chapters 3 and 5 of this volume for the influence of the media on this matter). In contrast, attention to terrorism increased exponentially once Rodríguez Zapatero won elections and Mariano Rajoy became the leader of the main opposition party. Similarly, as we explain in detail in Chapter 6, both the PSOE and especially the PP increase the number of questions about political decentralization when they are in the opposition as a strategy to force the government to take a position about one of the most controversial issues in Spanish politics. Again, this strategy is especially intense for the case of the PP. To talk about political decentralization is especially costly in electoral terms for the PSOE, among other things because most of its electorate opposes increasing decentralization, but the votes of the PSOE in Catalonia and the Basque Country are determinant to win the general elections. In contrast, the PP has a marginal position in both territories, which gives it more room to radicalize its policy position about terrorism and political decentralization.

In short, political parties select among the issues they want to politicize in the first place, changing their priorities depending on their governmental status. While in government, parties use oral questions as a means to give publicity about issues characterized by highly diffused costs and concentrated benefits for some collectives or territories. In contrast, as expected when in the opposition, both the PP and PSOE tend to concentrate attention on policy failures and controversial governmental decisions, which in some occasions are associated with unexpected events like an environmental catastrophe. This strategy of attacking the enemy is especially relevant for the case of crime and justice and governmental issues. In both cases the number of oral questions increase exponentially when the PP and PSOE are in the opposition, as a strategy to force the governing party to take a position about highly controversial or unpopular issues. In this respect, issue attention is a conditional phenomenon, by which parliamentary groups not only try to reinforce issue ownership, but especially to embarrass the governing party in order to maximize electoral rewards.

Conclusions

This chapter has illustrated how the government-opposition game is played in the parliamentary arena. We demonstrated that reforms in the formal rules that govern the introduction of questions in the Chamber affect agenda fragmentation and the use of different types of questions, and also create incentives for the governing and opposition parties to be more or less active during parliamentary oversight sessions. Results illustrate political parties have no interest to increase non-parliamentary activities. On the contrary, changing rules in 2008 altered a longstanding dynamic in which both the governing and main opposition party asked almost the same average number of questions per session, with a new situation in which the governing party has no incentive to maintain such oversight activity. As a result the number of oral questions declined across the last years, contradicting existing analysis about issue ownership and parliamentary activities.

Also, we explain how, in order to maximize electoral rewards, parties strategically select the type of question, renouncing non-routine questions when they are governing and increasing the number of urgent questions when they are in opposition. Similarly, in order to optimize parliamentary oversight sessions as a tool to damage the government, opposition groups tend to introduce more oral questions when the public perception of the economic and political situation is bad. This rational behavior also results in significant variations in the type of issues that are emphasized in the Chamber over time. In the framework of the constraints imposed by institutional factors, parliamentary groups strategically select the issues they want to politicize, focusing on those where there is not an agreement and that can deliver more electoral rewards. As we have shown, issues like terrorism and political decentralization are only emphasized by political parties when they are in opposition, as they want to force the government to take a clear position on an issue that may have important electoral consequences. Variations in issue competition are thus explained by party preferences but also by other variables like governmental status, variations in public opinion, or even prior attention to the issue. In the next chapter, we analyze to what extent the media influences parliamentary activities, contributing to the politicization of certain issues.

5
Media and Politics in Spain

Spanish politics play out in the media. Political elites seek media access to promote new ideas and policy proposals, attack their political adversaries, and defend their party manifestos during the electoral campaign. They seek access because the attention derived from it allows those elites to exist or be present in the public mind (Castells, 2009; Iyengar and McGrady, 2007). The media has increasingly become a venue where political parties compete with each other in order to have influence over public policy: It has become a "space of power making" (Castells, 2009; Strömback and Van Aelst, 2013). While political elites may view media as a tool to be used, the media also follows its own logic and pursues its own goals that are independent from the routines and requirements of governmental actors and political parties. Understanding how political actors use the media and are portrayed in it demands attention to the dynamics and business logic of the media, why it covers politics, and what journalists and media outlets seek to gain by doing so.

A large group of scholars suggest that media groups downplay coverage of specific policy proposals and increasingly focus on soft news such as sports, the weather, celebrities, and highly controversial issues in order to maximize their market share (Hallin and Mancini, 2004; Strömback and Van Aelst, 2013). We will call this "mediatization theory" because it focuses on the commercial logic of coverage of politics, suggesting that as the political game becomes "mediatized," so also is it trivialized and sensationalized. The limited amount of attention allocated to politics focuses on dramatized events including corruption or other types of scandals, polling data, public perceptions of politics and policies, and the differences rather than the similarities of elites (Iyengar, 1991; Iyengar and McGrady 2007; Graber 2003). By default and design, this means that many important policy issues, including health reform, education, agriculture, or foreign affairs, are ignored. Media outlets and sources do this because soft news is more commercially beneficial, and these "hard news" items may be dry and unappealing to the marginal reader. Mediatization theory is quite a pessimistic one with regards to the value of the media in enhancing vigorous or reasoned public debate.

"Indexing theory" provides a more positive assessment of the role of the media in covering politics, though it is not particularly flattering toward individual journalists (Bennet, 1990). According to indexing theory, journalists focus on what the government does – government's debates and policy proposals. Additionally, they place and emphasis on "official" sources, discount opinions that are marginal or not represented in the halls of government, and give priority to the routine reporting of government actions. Bennett's indexing idea comes from the observation that journalists often cover "beats" (such as the Prime Minister, Parliament, or a particular political party), that these "official" sources are the routine and reliable sources of useable information and quotations/reactions to the events of the day, that readers want to see "official" actions reported in the news, and that it is much easier to simply respond to and report what the government elites are doing than to go out and find stories about health reform, transportation problems, or other issues of government independently from government leaders. It is much easier simply to report what they do (or in the extreme, to reprint the press release coming from a government official).

Well-known trends toward the commercialization of the media, increased globalization, and greater competition from new media sources might make the indexing theory seem quaint and outdated. Certainly, one might argue, mediatization has won out, or will soon. But changing market conditions may actually promote dependency of media groups on governmental actors for several reasons. First, governmental actors are the main regulators of media markets and advertising practices. Second, they are a legitimate source of information. Third, government actors and the general political elite are relatively inexpensive sources of news; they are easy to access and provide potential quotations through press releases or written reports (Castells, 2009; Wolfsfeld and Sheafer, 2006). These characteristics are especially relevant as competition increases, resulting in decreasing market shares and potentially fewer active journalists in the newsroom and on the beat. As a result, media companies have reduced the set of resources devoted to direct coverage of news, both in terms of capital and the number of bodies on the ground. This in turn has reduced their capacity to provide reliable and independent information (Curran et al., 2009). All this implies that market forces and the mediatization of political news may reinforce rather than attenuate the factors already identified in indexing theories, which give advantage to "official" sources of the news.

Both approaches consider media and politics to constitute two independent systems guided by different logics. Both stress a unidirectional, exogenous, relationship linking journalists and policymakers, but they differ greatly on their assumptions about who is influencing whom (Strömback and Van Aelst, 2013). In contrast, agenda-setting scholars suggest the relationship between the media and politics is bi-directional and contingent (Walgrave and Van Aelst, 2006). Policymakers and journalists interact differently depending on the type of issues, the electoral cycle, and the institutional position of political elites in the government-opposition game (Green-Pedersen and Stubager, 2010; Thesen, 2013; Baumgartner and Chaqués-Bonafont, 2015).

In this chapter, we follow the agenda-setting approach. We expect that Spanish journalists will follow the ongoing political debate, prioritizing those issues that generate controversy between the Spanish government and opposition parties. In doing so, media groups follow their own market logic – routine issues are no news – but also respond to the needs and preferences of their allied political parties. The Spanish media system is one of the most highly polarized, with each party clearly allied with one of the major news outlets (see Chaqués-Bonafont and Baumgartner, 2013; Palau and Davesa, 2014; Baumgartner and Chaqués-Bonafont, 2015), so the journalists are attuned to their allied party leaders in different ways than the rival parties, and this is particularly true when one of these controls the government. Agenda scarcity leads parties to concentrate their efforts to gain media access on issues that either present a negative image of the governing party or to reinforce issue ownership (Green-Pedersen and Stubager, 2010; Thesen, 2013). Newspapers will contribute to this issue-selection process, but not in a neutral way. They will try to maximize the benefits of their allied party by highlighting the failures and inconsistences of the opposing party in relation to highly politicized issues, focusing on what is wrong, especially when the government is controlled by the rival partisan camp (Baumgartner and Chaqués-Bonafont, 2015).

Our results illustrate globalization and market changes did not result in a decline of media coverage of "hard news" politics. Our analysis of all the stories on the front pages of the two leading Spanish newspapers, *El País* and *El Mundo*, shows that the percentage of soft-news stories has not increased from 1996 to 2011, that attention to political scandals follows a cyclical trend, and that media attention is always concentrated on a few topics. Additionally, we show that there is a high level of correlation between the two newspapers in terms of the stories or topics that appear on their respective front pages.

However, our results also illustrate that there are important differences in the way the two newspapers perform their tasks. First, attention to issues is more concentrated on just a few topics in *El Mundo*; this appears to be related to that newspaper's high coverage of political scandals and terrorism. Second, both newspapers pay significantly more attention to the political scandals of their opponents rather than their allies. Third, parliamentary activities tend to follow media coverage, especially for opposition parties. Results illustrate that both the PSOE and the PP follow the media with a larger intensity when they are not governing, with independence of the number of questions they are asking, and with some important differences across issues.

Finally, these results also indicate that the internationalization of media groups, increasing foreign investment, and audience competition have not dramatically altered media coverage of politics in Spain; nor did they change the basic institutional features of the Spanish media system (Hallin and Mancini, 2004). More than ever, the Spanish media system is characterized by a strong division of readers according to ideology, high state intervention, and a radical erosion of media freedom. This directly affects the way and intensity with which Spanish citizens become informed about the political, economic, and social context in which they are involved.

The Spanish media system

Spain perfectly fits into the polarized pluralist media system, which is characterized by high political parallelism, low circulation of newspapers, low professionalism of journalists, and high levels of state intervention (Hallin and Mancini, 2004; Seymour-Ure, 1974, 2003; Gunther and Mughan 2000). The strong link between media groups and political parties affects the capacity of journalists to report on political campaigns in a neutral manner (Sampedro and Pérez, 2008), the people's knowledge about politics (Fraile, 2010), media coverage of certain issues like political corruption (Chaqués-Bonafont and Baumgartner, 2013; Palau and Davesa, 2013), and the capacity of policy actors to garner attention (Seymour-Ure, 1974; Castells, 2009; Baumgartner and Chaqués-Bonafont, 2014). Thus, it is fair to say that coverage of politics is strongly affected by partisanship – at least for the two main daily newspapers – strongly attached to one or the other of the largest mainstream political parties.

Politicization of media outlets and the ideological fragmentation of readership have been in place since the transition to democracy (Gunther and Mughan, 2000). During the late 1970s and 1980s this link was especially strong between newspapers and political parties as in the linkages between *El País* and the PSOE, *ABC* and the PP (previously *Alianza Popular*), or *La Vanguardia* and Convergencia i Unió (CiU). In the case of *El Mundo*, most readers identify themselves as conservative, as we show in Table 5.1. Still, its role as watchdog of Spanish politics – especially of PSOE governments – has captured readers from many different ideological branches, including the Far Left in the 1990s (Gunther and Mughan, 2000). *El País* and *El Mundo* are consistently the two most widely read newspapers in Spain, followed by other conservative newspapers like *ABC* or *La Razón* and territorially based newspapers like *El Periodico de Catalunya, La Vanguardia,* or *El Correo Vasco.*[1] Table 5.1 gives a summary of the major public and private media organizations in Spain, their year of creation, and the ideological distribution of their audiences.

Table 5.1 makes clear that the strong "political parallelism" (different audiences turning to different newspapers, based on ideology) is also present in the television and radio markets – for example, the link between *La Cope* and the Catholic Church – and has not been eroded despite two waves of regulations in the 1980s and 2000s. The first regulatory wave started in 1983, after passing Law 46/1983 on Regulating Regional Public Television,[2] and was reinforced in 1988 after the passing of Law 10/1988 on Private Television,[3] which set the general legal framework of private TV channels, and later by the Law on Local Television,[4] passed in 1994. As a result, in the early 1990s, public television (RTVE) had lost its monopoly of power in favor of regional public television channels, two private TV channels (Antena 3 and Tele 5), and two pay-per-view channels (Canal Satélite Digital and Via Digital, merged in 2002 to form Canal Plus).

The second regulatory wave started in the late 1990s, with Law 12/1997 on Liberalization of Mass Communications,[5] later amended by Law 7/2010 on Audiovisual Communications.[6] These regulations adapted the functioning of media market to technological changes, allowing for the proliferation of a vast set of TV channels and radio stations under the strict control of governmental actors that regulate licenses, advertising, etc. Fragmentation did not imply more market competition, but rather the consolidation of a few media groups – Grupo Prisa, Unidad Editorial (RCS) Vocento, and Planeta, Grupo Godó, Grupo Zeta – which diversified their activities across media outlets (TV, radio, and newspapers). Also, this allowed for their increased presence in foreign markets, especially in Latin America, and it allowed for the entry of foreign capital. These reforms intensified the financial dependency of "local" outlets vis-à-vis foreign groups and financial institutions, such as Viacom, Time Warner, and News Corporation (Castells, 2009; Jones, 2007; Berges, 2010). This is the case of *El Mundo,* which integrated in

Table 5.1 Political parallelism of media groups in Spain

Type	Media group	Media outlet	Name	Year of creation	Ideology of audiences (2009) Left (1–2)	(3–4)	(5–6)	(7–8)	Right (9–10)
Private	Grupo Prisa	Newspaper	*El País*	1976	15.6	57.7	19.2	4.0	0.9
		Radio	La SER	1924	15.5	49.9	20.7	8.0	1.2
		Television	Canal Plus*	1993					
	Mediaset	Television	Tele 5	1990	6.7	30.3	31.9	13.1	3.7
		Television	Cuatro	2005	15.1	37.9	26.3	8.3	2.3
	Grupo Vocento	Newspaper	ABC	1903	2.5	5.0	33.1	43.8	11.6
		Television	Inter-economía	2005	0.0	4.9	21.8	43.5	22.9
	Conferencia Episcopal	Radio	La Cope	1960	2.6	7.8	34.5	37.2	9.9
	Unidad Editorial (RCS)	Newspaper	*El Mundo*	1989	2.5	13.8	40.1	33.9	5.0
	Grupo Zeta	Newspaper	El Periódico Catalunya	1978	17.1	31.0	37.2	10.1	2.3
	Mediapro		El Público	2007–2012	52.0	48	0.0	0.0	0.0
	AtresMedia (Planeta)	Newspaper	La Razón	1998	2.4	2.4	41.5	26.8	14.6
		Radio	Onda Cero	1989	3.8	21.1	37.4	25.5	4.6
		Television	La sexta	2006	18.3	43.9	23.2	6.7	1.2
		Television	Antena 3	1990	4.0	19.0	38.3	21.4	4.3
	Grupo Godó	Newspaper	La Vanguardia	1881	10.9	27.6	35.6	16.7	4.6
		Radio	RAC1	2000	11.7	33.0	43.6	3.2	2.1
		Television	8TV	2001					
Public	CCRTV (Catalonia)	Television	TV3	1983	17.6	31.6	31.6	9.3	1.6
		Radio	Catalunya Radio	1983	27.7	33.8	27.7	6.0	2.4
	RTVE	Television	TVE1	1956	9.5	34.3	28.7	12.5	3.3
		Television	TVE2	1965	10.9	37.8	16.2	18.9	0.0
		Radio	Radio Nacional de España		13.3	32.7	30.5	12.1	3.5

Note: *Canal Satélite Digital merged with Via Digital, pay-per-view.

Source: Ideology of readers comes from the *Centro de Investigaciones Sociológicas* (www.cis.es) survey number 2920/0, question numbers 25a, 25b, and 25c.

full in the Italian media group RCS. Another example is the merger of *Telecinco* owned by the Italian group Mediaset and Cuatro (Grupo Prisa) (Jones, 2007; McChesney, 2003).

Market competition did not limit political parallelism. As illustrated in Table 5.1, both a given news outlet and its viewers/readers are associated with a specific political ideology. More than 50 percent of the readers of *El País* are voters of the PSOE or IU and identify themselves in the Left or Far Left. Fully 23 percent of the readers of *El Mundo* identify themselves as PP voters, versus 4 percent and 2 percent of the voters of the PSOE and IU respectively[7] (CIS, 2009; Gunther and Mughan 2000). So there is very strong evidence of political parallelism, and the various consolidations and cross-sector mergers in the communications industry have not softened this aspect of Spanish politics and media.

Additionally, increasing market competition was not accompanied by larger audiences, especially in the case of newspapers. On the contrary, media audiences[8] and advertising revenue have declined due to increased competition for a static number of consumers. In some cases, such as *El País,* this has led to reducing salaries and cutting staff. In others, like *El Público,* this has resulted in bankruptcy. In the case of public media groups at the national and regional level, several reporters were removed after voicing criticism of the government's policies (Freedom House report, 2013). This has also affected the directors of some of the most relevant newspapers in Spain, like Pedro J. Ramírez, the director and cofounder of *El Mundo* from 1989, removed in January 2014 after an extensive campaign to highlight the scandals of corruption of political elites in and out of the Spanish government, including the Spanish crown. So not only is the Spanish media highly partisan in orientation, but it has also come under threat from the government itself.

The erosion of media freedom

The above description suggests that, contrary to the mediatization theory, globalization and market competition have not strengthened the capacity of journalists to set their own priorities. In fact, institutions like the Freedom House[9] or Journalists without Borders, demonstrate that since 2000, the ability of media outlets to operate freely and without fear of repercussions has decreased dramatically as a result of legal, political, and economic pressures.[10]

Traditionally, Spain has been among the countries with relatively high standards of media quality, but political and financial institutions increasingly restrict the ability of journalists to operate freely.

Political independence of public television has not increased despite the reforms of the public broadcasting service (Radio television Española, RTVE) introduced by the first government of Rodríguez Zapatero. On the contrary, the perception that RTVE was highly politicized gradually increased to the extent that the PP refused to consider RTVE for the televised electoral debates of the 2008 electoral campaign (Sampedro et al., 2008). Additionally, political control over the content of news media has increased in terms of censorship. More important perhaps than official actions in this area, their inherent threat has led to a substantial degree of self-censorship, limiting the ability of reporters to cover the news freely. Media coverage of electoral campaigns and internal meetings of political parties is controlled by the parties themselves. Parties maintain this control by banning journalist intervention and rationing who may take and use pictures and attend campaign events (Sampedro et al., 2008).

Van Dalen and Van Aelst's study (2014) on journalist perceptions of the media agenda-setting power in eight European democracies – Belgium, Denmark, Germany, Norway, Spain, Sweden, and the Netherlands – provides further evidence of this fact. In the late 2000s, more than two-thirds of Spanish political journalists stated that political elites have greater agenda-setting power than themselves; the reverse was true in many of the other countries such as in Sweden or Belgium (Van Dalen and Van Aelst, 2014: 11). None of the Spanish political journalists totally agreed with the question "the media decides which issues are important, politics has little impact on this matter," as compared to 9.3 percent of journalists in Sweden or 6.6 percent in Belgium and Norway.

In short, for the last decade there has been a gradual erosion of media freedom that occurred in parallel to globalization and growing market competition. Political elites have adapted to the way media outlets produce and disseminate information, their standards of newsworthiness, and the changing norms of the profession of journalism. These trends have resulted in reduced journalistic autonomy over time. From here, the next question is whether political and market changes have altered media coverage of politics. This broad question also breaks down to a series of questions such as whether the front pages are increasingly dominated by soft news, political scandals, and horse-race journalism, and/or to what extent media attention is increasingly concentrated in a few issues. We turn to these questions in the next section before turning, at the end of the chapter, to ask whether the media follows the ongoing political debate, and to what extent policymakers are attentive to media attention.

Media coverage of politics

From 1996 to 2011, approximately 18 percent of the stories in the front pages of *El País* and *El Mundo* were related to soft news. In the case of *El País*, attention to soft news focuses on culture and celebrities, which accounts for about 55 percent of all soft-news stories, and sports news accounted for about 40 percent of all soft-news stories. In the case of *El Mundo*, soft news is equally distributed between sports and culture – about 40 percent each. Weather and death notices cover the rest of soft news. Figure 5.1 shows these trends.

Soft news, that is, news with no policy content, fluctuates substantially depending on major sporting events that can occasionally garner over

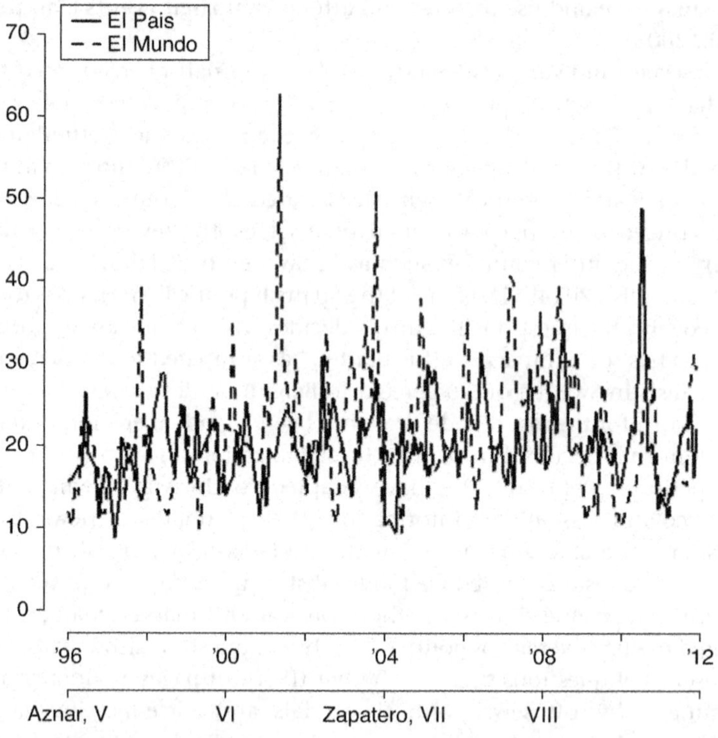

Figure 5.1 Percentage of stories without political content

Note: The figure shows the percentage of stories on the front pages of each newspaper, by month, that have no political or public policy content. Such "soft news" stories include sports, entertainment, and other topics unrelated to government, politics, or public policy.

Source: Spanish Policy Agendas Databases (www.ub.edu/spanishpolicyagendas).

50 percent of all front-page news. However, while the number of such stories varies from month to month by substantial amounts, there is no strong trend toward a crowding-out effect whereby soft news consumes an ever-increasing portion of the available news hole, at least on the front pages of these two major newspapers.

Figure 5.2 illustrates that media attention to political scandals follows a cyclical pattern. Attention is especially high in the first legislature governed by José M. Aznar (1996 to 2000) – more than one story every two days – and later during the PSOE government of José Luís Rodríguez Zapatero (2008–2011) – an average of 3 percent and 7 percent for *El País* and *El Mundo*, respectively. In the 1990s, political scandals were mostly linked to the PSOE (like the Caso Roldan and the GAL scandal), while in the mid-2000s when the PP were in government cases of corruption

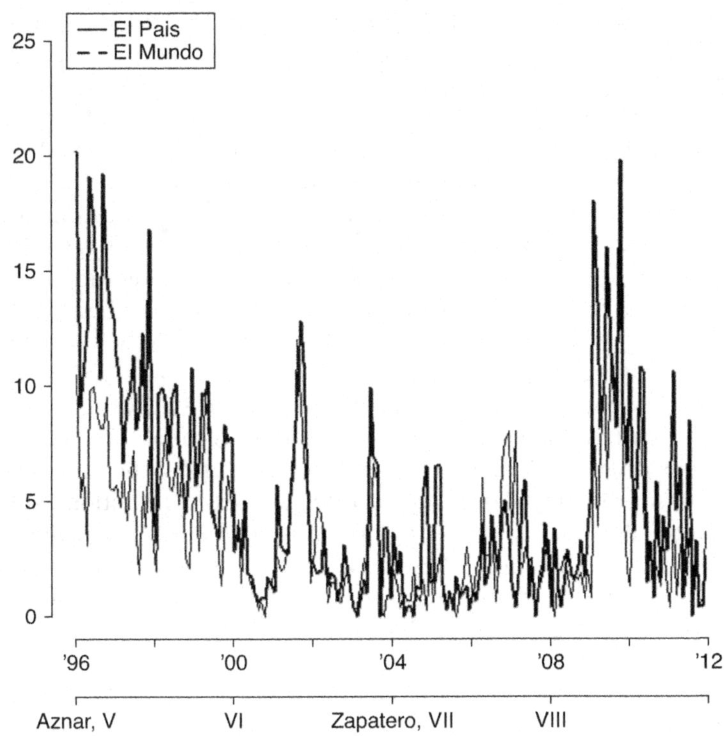

Figure 5.2 Percentage of stories about political scandals

Note: The figure shows the percentage of stories on the front pages of each newspaper, by month, on the topic of corruption.

Source: Spanish Policy Agendas Databases (www.ub.edu/spanishpolicyagendas).

were more and more associated with the PP, especially in Valencia (Caso Correa) and the Comunidad Balear (Caso Matas).

Despite this, as Table 5.2 illustrates, newspaper readers may get a different picture of political scandals in Spain depending on which newspaper they read. In fact, media coverage of corruption scandals is always significantly higher in *El Mundo*, except when the scandals relate to that newspaper's allied party, the PP, in which case coverage is higher in *El País*. In the late 1990s, both newspapers reported about cases that directly eroded the prestige of the PSOE, including the GAL scandal, and Caso Roldan (about 6 percent of the stories) without significant differences across the two newspapers. However, in the same period, *El País* reported double the number of stories as *El Mundo* regarding the PP. These differences were even larger in the next legislature, were *El País* reported five times more than *El Mundo* about PP scandals. Since 2004, the situation reversed. Both newspapers paid similar attention to the political scandals of the PP (about 6 percent for both, from 2004 to 2008), but in the same period, *El Mundo* devoted 4.7 percent to front-page stories about the PSOE, versus 0.35 percent in *El País*. Since 2008, differences in media coverage increased even more; *El Mundo* devoted 14 percent to PSOE scandals, which is more than one story a day, while most of these scandals are almost ignored in the *El País* front pages. So clearly, coverage of scandals relating to political leaders is related to whether those leaders are currently in power, and whether they are from an allied or a rival ideological camp to the newspaper in question.

Figure 5.3 illustrates that attention to polls, public perceptions of politics, and the organization of the political parties follows a cyclical trend in both newspapers. Increasing attention is not necessarily linked to elections – attention increases right before or after electoral campaign – but also between elections due to a variety of factors, from the organization of the first primaries of the PSOE in 1998, to the confrontation between Mariano Rajoy and Esperanza Aguirre after the PP lost the majority in the 2008 general election. These results do not give support to the idea of growing horse-race journalism in Spain; there is no trend over time toward increased coverage of these matters. Rather, they come and go depending on events.

Finally, we ask whether attention is increasingly concentrated on a few issues, especially on those that are more politicized. Chaqués-Bonafont and Baumgartner (2013) demonstrate that media attention is highly correlated; both newspapers overlapped in the issues they pay attention to. However, they do so with different intensities. The number of stories about the economy, foreign affairs, or any of the welfare-related issues,

Table 5.2 Differences across newspapers in attention to political scandals, soft news, and the horse race

	Legislature							
	1996–2000		2000–2004		2004–2008		2008–2011	
	Mean difference	Sig. (t-test)	Mean difference	Sig. (t-test)	Mean difference	Sig. (t-test)	Mean difference	Sig. (t-test)
Number of stories, *El Mundo-El País*	–17.000	.000	–8.000	.000	–7.063	.000	-9.227	.000
% Stories on political corruption, *El País-El Mundo*	–3.792	.000	–.536	.028	.096	.776	-2.153	.000
% Soft news, *El País-El Mundo*	–.413	.540	–1.082	.140	–.985	.266	-1.301	.09
% Stories PP political corruption, *El País-El Mundo*	18.571	.000	31.392	.000	35.433	.000	43.237	.000
% Stories PSOE political corruption, *El País-El Mundo*	–12.833	.000	–18.240	.000	–35.570	.000	-32.453	.000
% Difference in coverage of PP corruption minus PSOE corruption, *El País*	–21.614	.002	41.649	.000	33.850	.000	74.859	.000
% Difference in coverage of PP corruption minus PSOE corruption, *El Mundo*	–53.018	.000	–7.983	.200	–37.154	.000	-.831	.912
% Horse race, *El País-El Mundo*	–.865	.069	–3.571	.001	–3.095	.013	-4.27	.000

Note: All calculations are at a monthly level. “Horse race” refers to discussions of which party is leading or trailing in the electoral campaign.

Source: Spanish Policy Agendas Databases (www.ub.edu/spanishpolicyagendas).

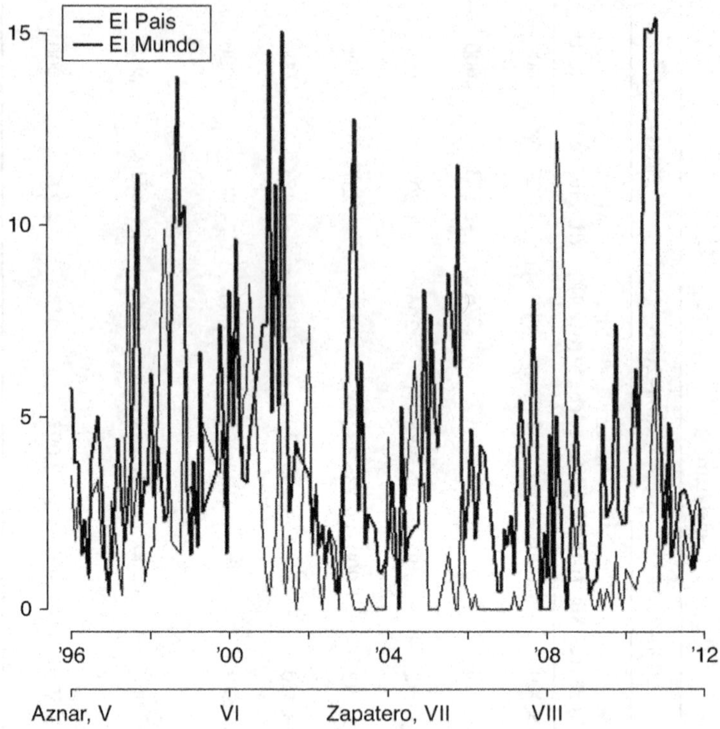

Figure 5.3 Percentage of stories about the electoral horse race

Source: Spanish Policy Agendas Databases (www.ub.edu/spanishpolicyagendas)*Note*: The figure shows the percentage of stories on the front pages of each newspaper, by month, focusing on the "horse race" aspect of elections and party competition.

from labor to education is always larger in *El País* than *El Mundo*. The two subtopic exceptions to this are coverage of political scandals and terrorism. *El Mundo* pays more attention to these two issues resulting in a more concentrated agenda for the whole period, as illustrated in Table 5.3 and Figure 5.4. Note that newspapers have the same agenda capacity, about eight stories each front page, and results are equally significant if we take into account the absolute percentages. As we argue in the next section, this illustrates the willingness of *El Mundo* to perform the role of watchdog, and to follow its own agenda independently of the parliamentary debate.

In short, results so far indicate that media outlets respond to their audiences by providing a one-sided view of the political debate and paying a disproportional share of attention to their opponents. Both *El País* and

Table 5.3 Difference in attention across issues in *El País* and *El Mundo* during election and non-election periods

Topic	All periods	Sig. (t-test)	Non-electoral periods	Sig. (t-test)	Election periods	Sig (t-test)
Economy	.978	.000	3.395	.000	.750	.681
Rights	.936	.000	3.587	.000	8.833	.030
Health	.309	.018	1.756	.001	.500	.747
Agriculture	.245	.001	.802	.036	.333	.707
Labor	.208	.129	1.046	.032	1.250	.420
Education	.186	.297	1.227	.029	2.666	.187
Environment	.563	.000	1.535	.004	1.333	.088
Energy	.298	.003	1.273	.007	.166	.902
Transportation	.631	.000	2.163	.000	3.75	.081
Crime & justice	−5.081	.000	−6.651	.000	−3.833	.524
Social policy	.275	.000	.930	.000	.666	.399
Housing	.452	.000	1.378	.000	.250	.778
Commerce	.563	.002	2.581	.000	−.583	.796
Defense	1.001	.003	4.070	.011	2.750	.427
Research & development	.448	.017	2.273	.000	2.083	.212
Foreign trade	.317	.000	0.907	.002	1.000	.084
International affairs	1.716	.000	5.523	.000	6.833	.012
Government	−4.247	.000	−3.645	.000	−9.583	.139
Public lands	.204	.000	.517	.023	1.250	.012
Terrorism	−.030	.000				

Note: Entries show the difference in percentage attention to each topic in *El País* minus that in *El Mundo*. Election periods are defined as the three months before elections. Data are monthly from 1996 through 2011.

Source: Spanish Policy Agendas Databases (www.ub.edu/spanishpolicyagendas).

El Mundo highlight the faults of their enemies more than the virtues of their allies, focusing on who is responsible for an unsolved problem or a controversial decision (Baumgartner et al., 1997; Herman and Chomsky, 1988; McChesney, 2003; Baumgartner and Chaqués-Bonafont, 2014). In highlighting the weaknesses of their political opponents, both newspapers seek to persuade or even manipulate public opinion by taking into account the preferences of their allies (Palau and Davesa, 2013; Baumgartner and Chaqués-Bonafont, 2014). This all serves to allay the fears that may follow from some theories about the mediatization of politics discussed earlier in this chapter; attention is not increasingly focused on soft news. Additionally, the coverage of political scandals or political parties as organizations follow a cyclical process, reflecting the changes in the political context, the electoral cycle, and focusing events

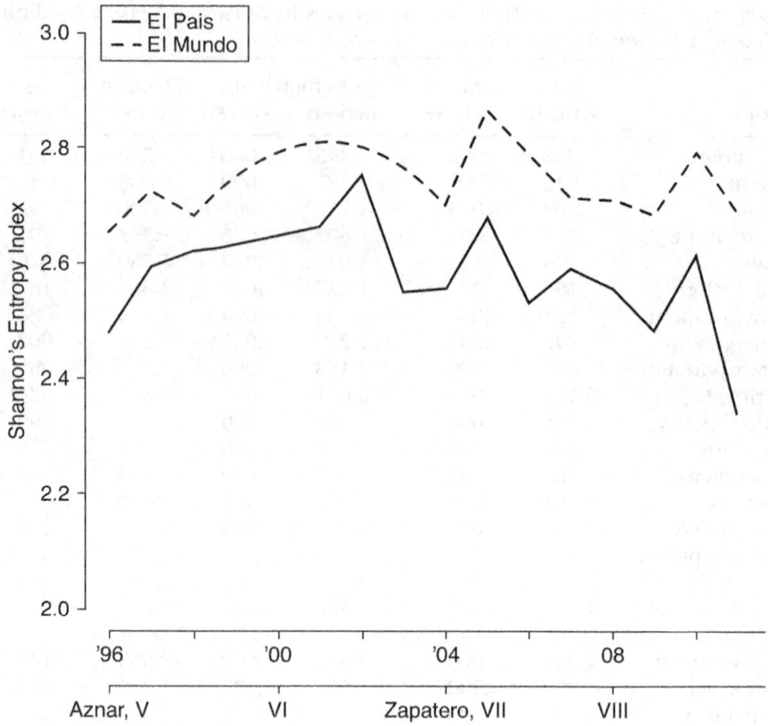

Figure 5.4 Fragmentation of the media agenda

Note: The figure shows Shannon's entropy index, a measure of diversity of attention. Both newspapers have similar values over time, though *El País* is consistently slightly broader in its coverage.

Source: Spanish Policy Agendas Databases (www.ub.edu/spanishpolicyagendas).

like a new case of corruption. In sum, while there are some differences by party in their coverage of politics, especially bad news affecting their rivals, there are few strong trends over time suggesting an abandonment of "hard" news in favor of entertainment, horse-race issues, or other stories without policy content.

Parliamentary activities and agenda setting by the media

While we have demonstrated that media tracks politics, the question still remains whether parliamentary activities follow media attention. In a democratic polity, one expects the media to inform the citizenry about politics, which should give voice to a broad set of policy actors. In performing these tasks, the media selects among issues, which it hopes will maximize its goals. While doing this, the media takes into account the strategies of the political elites. Issues often get onto the media agenda as a result of a routine process relating to government activities; for example, every Friday, the media covers the decisions taken by the Council of Ministers. Others arrive as a result of a sudden change or focusing event. In many other cases, newspapers reflect the efforts taken by political elites to politicize some issues. In doing so, the media reinforces the skew and explosiveness that characterizes changes in the allocation of attention by governmental actors (Chaqués-Bonafont and Baumgartner, 2013).

As in the case of parliamentary and governmental agendas, media attention is highly concentrated in a few topics. For a long period of time, political elites, the media, and citizens in general pay almost no attention to most problems, regardless of their severity. These periods of inaction and status quo are broken by dramatic change that leads to overreactions and bursts of attention to some issues (see Boydstun, 2013, for similar findings about the US *New York Times*). From here, the question still stands as to whether political parties react differently to media attention depending on their preferences and their structural position in the political system (Green-Pedersen and Stubager, 2010; Thesen, 2011; Vliegenthart and Walgrave, 2009,; Baumgartner and Chaqués-Bonafont, 2015). In a context of agenda scarcity, political parties focus their efforts on politicizing those issues that respond better to their interests and maximize their possibilities of winning in the next election. While in government, parties publicize their policy programs and achievements across a broad range of issues – especially those that generate no political conflict but are nonetheless relevant for Spanish citizens. Examples of these policies are a new program to fight against drugs, or the creation of a national regulatory agency to fight against corruption. In contrast, parties while in opposition focus their attention on highly controversial issues, stressing the problems and mistakes associated with governmental decisions and/or unexpected events, like food safety or environmental catastrophes. When introducing oral questions, opposition parties emphasize those aspects of the governmental agenda that are more controversial, leaving aside those issues for which there is a general agreement. So, opposition parties can pick and choose among those issues that may make the government look bad, but the government parties are forced to discuss a broad range of issues, even those they do not like, and even as they try to bring attention to those issues on which there is a consensus and for which they seek to gain credit among the public.

In this process of issue selection, the media and opposition parties reinforce each other by highlighting negative news about politics and the government. Journalistic norms oblige them to focus attention on government actions, as these are seen as inherently newsworthy. This is especially true for highly politicized issues (i.e., those issues involving high degrees of conflict), because political leaders are eager to participate in expressing their views using a populist discourse (Iyengar and McGrady, 2007; Jones and Baumgartner, 2005; Graber, 2003). In the search for securing an audience share, newspapers of all ideological lines might focus on topics like political decentralization, rising unemployment, or the environment independently of which political party "owns" the issue (Baumgartner and Chaqués-Bonafont, 2013). Rather, journalistic norms and media-functional logics limit the capacity of media outlets to contribute to the politicization of some issues and not others, according to the preferences of their partisan allies.

Analysis

Taking all these arguments into consideration, we test to what extent the percentage of oral questions asked by the PP and PSOE devoted to a particular issue by the governing or the main opposition party each month (t) is affected by the percentage of media attention in a newspaper the previous month (t–1). Results of a simple OLS regression are summarized in Table 5.4.

As expected, political parties have a higher probability of following the media when in opposition than when in government. Actually, while in government correspondence between the oral questions and the stories in the front pages of *El País* and *El Mundo* is almost inexistent for all issues with the exception of labor and environmental issues in the case of the PP, and defense in the case of the PSOE (here only for the case of *El País*). The probability that the deputies of the PSOE ask questions about defense issues increases especially after Rodríguez Zapatero wins elections in 2004, and takes the decision to end the participation of Spain in the war of Iraq, following one of his most visible electoral promises. Media stories in the front pages of *El País* reinforce the willingness of PSOE deputies to highlight one of the milestones of the Rodríguez Zapatero government that had the support of more than 80 percent of Spanish citizens, and that generated a vast set of demonstrations all over the country. By the same token, deputies of the PP increased their attention to labor-related issues, in a context of economic success in which

Table 5.4 Explaining the impact of the media on parliamentary activities

Topic		El País						El Mundo					
		PSOE			PP			PSOE			PP		
		B	Sig.	R^2	B	Sig.	R^2	B	Sig.	R^2	B	Sig.	R^2
Economy	Opposition	1.241	.001	.144	2.471	.000	.343	1.788	.000	.220	2.795	.000	.390
	Government	–.012	.954	.000	.024	.935	.000	.159	.449	.010	–.033	.926	.000
Rights	Opposition	–.066	.702	.002	.127	.644	.004	–.057	.749	.001	.704	.003	.138
	Government	–.170	.578	.005	–.019	.834	.001	–.111	.687	.003	–.033	.731	.002
Health	Opposition	–.446	.422	.010	–.095	.567	.005	.124	.822	.001	.131	.385	.012
	Government	.490	.580	.005	–1.162	.011	.094	.430	.595	.004	–.183	.692	.002
Agriculture	Opposition	1.757	.000	.227	.709	.076	.048	.194	.000	.322	1.928	047	.059
	Government	.066	.941	.000	-.013	.979	.000	1.537	.477	.008	.520	.309	.015
Labor	Opposition	1.731	.012	.091	.090	.691	.002	.737	.205	.024	.399	.100	.041
	Government	–.303	.195	.026	1.040	.013	.088	– .048	.852	.001	1.231	.000	.178
Education	Opposition	1.370	.032	.067	–.451	.138	.034	1.216	.030	.069	–.161	.593	.004
	Government	.436	.478	.008	-.500	.349	.013	.584	.333	.014	.221	.636	.003
Environment	Opposition	1.022	.046	.058	-.017	.925	.000	.645	.189	.026	.106	.685	.003
	Government	–.758	.185	.027	.642	.007	.102	–.097	.907	.000	.565	.013	.088
Energy	Opposition	1.739	.001	.156	.698	.036	.066	1.479	.000	.175	.946	.001	.152
	Government	–.038	.861	.000	–.216	.411	.010	–.023	.907	.000	–.298	.156	.030
Transportation	Opposition	.755	.074	.047	.421	.284	.018	.428	.369	.012	.413	.260	.019
	Government	–.395	.200	.025	.253	.676	.003	–.170	.557	.005	1.063	.112	.037

Continued

Table 5.4 Continued

		El País						El Mundo					
		PSOE			PP			PSOE			PP		
Topic		B	Sig.	R^2	B	Sig.	R^2	B	Sig.	R^2	B	Sig.	R^2
Internal affairs	Opposition	–.028	.923	.000	.632	.035	.066	.188	.497	.007	.560	.015	.088
	Government	–.133	.520	.006	.069	.630	.003	–.171	.280	.018	.105	.431	.009
Social policy	Opposition	.133	.807	.001	.739	.154	.031	–.268	.548	.005	1.274	.106	.040
	Government	–.683	.361	.013	.244	.733	.002	–.205	.857	.001	.003	.997	.000
Housing	Opposition	4.446	.003	.125	.275	.336	.014	–0.131	.927	.000	.116	.838	.001
	Government	.249	.561	.005	.299	.709	.002	1.518	.070	.049	–.640	.390	.011
Commerce	Opposition	.923	.008	.101	.381	.207	.024	.390	.331	.014	.203	.426	.010
	Government	.069	.750	.002	–.037	.870	.000	–.091	.618	.004	–.267	.294	.016
Defense	Opposition	1.120	.000	.333	–.162	.195	.026	.780	.000	.510	.031	.869	.000
	Government	.389	.021	.080	–.085	.592	.004	.573	.022	.078	–.044	.626	.004
Research &	Opposition	1.200	.001	.144	.108	.554	.005	.397	.121	.036	–.125	.511	.007
development	Government	–.629	.272	.019	–.492	.212	.023	–1.931	.001	.160	–.475	.068	.049
Foreign trade	Opposition	–.065	.639	.003	–.089	.522	.006	–.054	.732	.002	–.138	.389	.011
	Government	.283	.317	.015	–.195	.480	.007	.428	.188	.027	–.506	.104	.039
International affairs	Opposition	.242	.311	.015	.534	.041	.062	.144	.518	.006	.188	.475	.008
	Government	.264	.485	.008	–.023	.944	.000	.315	.399	.011	.315	.290	.017
Government	Opposition	.374	.106	.038	.077	.807	.001	.318	.070	.048	–.063	.818	.001
	Government	–.208	.126	.036	.162	.114	.037	–.168	.153	.031	.051	.518	.006
Public lands	Opposition	.206	.811	.001	.187	.747	.002	–.303	.729	.002	–.183	.783	.001
	Government	1.610	.131	.035	3.772	.004	.118	-.617	.619	.004	2.624	.052	.055

Note: Entries are the results of OLS regressions where the dependent variable is the percentage of oral questions introduced by each party on the topics indicated, and the independent variable is the percentage of stories on the front page of the newspaper indicated in the previous month. Models are shown separately for periods when each party serves in government versus in opposition. Models also include a constant term, not shown. Data are monthly and cover the period of 1996 through 2011.

unemployment decreases to almost 10 percent for the first time since the transition to democracy. To achieve this goal, the PP put into practice a highly controversial labor market reform, generating an intense reaction by opposition parties.

However, these results illustrate that when asking oral questions in Parliament, members of the party in government do not follow the media. As we explained in Chapter 4, and in contrast to other advanced democracies, this lack of correspondence between stories and the oral questions of the governing party are not explained by the fact that MPs of the governing party do not ask oral questions. Thus the lack of correspondence cannot be related to the fact that governing parties ask fewer questions, but rather to the fact that governing parties attempt to shed light on those tasks and policies where the government does well. Good news and policy-intensive news have no special interest for the media, and thus they are less likely to cover it (Baumgartner et al., 2009).

In contrast, results in Table 5.4 illustrate deputies of the PP and PSOE tend to follow media stories of *El País* and *El Mundo* across a wide range of issues while in the opposition. First, both parties tend to follow both newspapers in relation to macroeconomics, agriculture, and energy regulations. The probability that the PP and PSOE ask an oral question about the economy increases when *El País* or *El Mundo* write about this issue, and this correspondence is especially important for the case of the PP and *El Mundo*. Actually, the robustness of the model in relation to macroeconomics is the highest. Additionally, these results illustrate there is no link between issues related to welfare generally and parliamentary activities, with the exception of education and only for the case of the PSOE. The more the stories about education policy, the more the questions asked by PSOE deputies about this issue, but only when the PSOE is in opposition.

Education shows itself to be the main exception, because it is linked with two of the most important cleavages in Spanish politics today: The role of the Catholic Church in education (and moral issues in general) and political autonomy of regional governments concerning education policy, and more specifically the use of languages in the curricula. On the one hand, the educational reforms of conservatives and socialists have always been aimed at modifying the role of the Catholic Church in the educational system. The socialists attempted to promote a radical change aimed at consolidating Spain as a nonconfessional state, autonomous from the Catholic Church, eliminating religion from the school curricula. In fact, education reform was one in a set of regulatory reforms, similar to changes to the regulations of same-sex marriage and abortion. that

Table 5.5 Comparing media attention in *El País* and *El Mundo* during election and non-election periods

	El Mundo		El País	
Topic	**Mean differences**	**Sig. (t-test)**	**Mean differences**	**Sig. (t-test)**
Economy	−.634	.398	.597	.221
Rights	1.308	.267	−1.141	.449
Health	.121	.758	.575	.320
Agriculture	.173	.562	.320	.253
Labor	.231	.655	.121	.807
Education	.547	.406	-.182	.788
Environment	.336	.111	.419	.205
Energy	.580	.192	1.058	.015
Transportation	.543	.149	−.090	.898
Crime & justice	.920	.593	−.933	.544
Social policy	−.238	.393	−.156	.533
Housing	−.094	.674	.348	.342
Commerce	−.748	.352	.683	.261
Defense	1.538	.112	1.494	.354
Research & development	−.047	.934	−.291	.549
Foreign trade	.502		.417	.092
International affairs	1.348	.058	.589	.520
Government	−6.444	.000	−3.529	.046
Public lands	.056	.538	−.296	.115

Note: Entries show the difference in attention from non-election times to during the periods of elections, defined as three months preceding a national election. Data cover the period of 1996 through 2011.

Source: Spanish Policy Agendas Databases (www.ub.edu/spanishpolicyagendas).

directly attacked the moral principles defended by the Catholic Church. The governmental agenda of Mariano Rajoy, after winning elections in 2011, illustrates the intensity of this political conflict. The debate about the recently passed education law (Ley Wert) was not centered on how to improve education standards in Spain, especially on the inclusion of religion in the school curricula, and the use of Catalan in the schools of Catalonia. A similar process is apparent with the new governmental bill about abortion, whose first draft revokes most of the changes introduced by the socialist government in 2010, as we explain in Chapter 8.

These conflicts have captured the front pages of *El País* and *El Mundo* for the last decades, reinforcing the already important polarization of the political debate in Spain. However, stories about education policy are only followed by PSOE deputies while in the opposition – the probability that the PSOE asks an oral question about education increases by 1.3 and 1.2 after greater coverage of the issue in *El País* and *El Mundo*, respectively. Coverage of education provides the opportunity for the PSOE – while in opposition – to politicize an issue especially rewarding in electoral terms. The contrary occurs for the case of terrorism: correspondence between oral questions and the stories in the front pages only exist for the case of the PP while in the opposition. In fact, the PSOE asked no questions about terrorism from 2000 to 2004 as a result of the *"Pacto por las libertades"* reached in the early 2000s, by which the PSOE took the compromise to avoid the politicization of terrorism for electoral purposes (see Chapter 3). Once the PSOE won the election in 2004, terrorism became one of the key issues in the parliamentary agenda of the main party in the opposition led by Mariano Rajoy.

Results indicate the PP followed media attention to raise the visibility of terrorism as an issue. The odds that the PP would ask an oral question about terrorism increase when *El País* or *El Mundo* report about this issue in their front pages in the previous week. This new politicization of terrorism took place in a context in which the number of victims had declined to almost zero and stayed at that level for an extended period. Results also indicate the PP tends to follow the media with more intensity in relation to rights issues, although results are only significant for the case of *El Mundo*. This is very much related with media coverage of immigration and moral issues (see also Chapter 8). This is not surprising taking into account that most parts of the drama associated to the arrival of immigrants to the coast of Spain occurred in the mid-2000s, right after the PSOE won elections. In short, these results illustrate that correspondence between parliamentary activities and the stories in the front pages of the two major newspapers in Spain are especially intense when parties are not governing, and for a small set of issues. Parties follow the media in relation to macroeconomics, unexpected events, and focusing events like agriculture (e.g., the mad-cow scandal), immigration, or environmental catastrophes; or in relation to those issues which are especially controversial and directly linked with some of the old cleavages in Spanish politics, like the role of the Catholic Church in relation to education policy and moral issues, terrorism, or as we will explain in the next chapter, political decentralization. In this regard, while in the opposition, the PSOE follows the media in relation to education, while the PP tends to follow the media with special intensity in relation to terrorism, rights, and political decentralization.

The different patterns of behavior between the two main parties illustrate the importance of the media as a new platform in which political conflict takes place. Both newspapers pay attention to a similar set of issues, with high levels of correlation. In both newspapers, political scandals, terrorism, immigration, political decentralization, and/or the war of Iraq are among the top most important issues covered in the front pages. However, in both cases, information is biased and more focused on the faults of their enemies than the virtues of their allies, which contributes to fostering confrontation among political forces. Authors such as Sampedro (2004) and Sampedro and Seoane Pérez (2008) arrive at the conclusion that these changes have generated a transformation of the Spanish media system, from a polarized pluralism to antagonistic bipolarization in which the two main statewide political parties increasingly dominate the political debate.

Political campaigns

We are now left with one final question: to what extent does media coverage of politics vary between routine times and during the period before elections? (We define the period of elections as three months before elections.) Table 5.5 summarizes differences in media attention between *El País* and *El Mundo* comparing routine and electoral times. Results clearly illustrate that whatever differences appear between the two newspapers are only significant during routine times, but during electoral campaigns these differences disappear. During routine times mean differences are significant for all issues and follow the same pattern as illustrated in the previous sections (Table 5.5). *El Mundo* always pays more attention to governmental issues (mainly political scandals and political decentralization) and crime and justice (mainly terrorism). Coefficients follow the same pattern during election time but are not significant anymore. This is the case for all elections without major differences across time. Also, as one may expect, attention to soft news decreases significantly during elections – to 14 percent and 13 percent in *El Mundo* and *El País*, respectively. Finally, mean differences in the attention to political scandals decrease significantly for both newspapers, including the cases in which they refer only to their opponents. This means media coverage of issues becomes more similar during election times, and more focused on the debate about policy issues and political campaigns instead of soft news.

Another question is whether issue attention within newspapers varies between election and routine times. Results in Table 5.5 illustrate attention to governmental issues increases significantly during elections, for both newspapers. This is not surprising as governmental issues include all stories about political campaigns and internal affairs of political parties. In fact, it is during elections when political parties gain more visibility in the front pages (Baumgartner and Chaqués-Bonafont, 2014). Parties gain newsworthiness, becoming an important source of information, supplying journalists with an overwhelming amount of easy-to-get news in different formats – from public events and ceremonies, a multitude of photo opportunities, reports and briefings about issues, provocative statements, and/or press releases. They become the protagonist of the political debate, and in doing so they search access to the media either to defend their party manifesto, to give visibility to the issues they own, or simply to attack their enemies on highly controversial issues.

Mean differences between routine and election time are not significant for any other issues, which basically means newspapers do not change the way they perform their functions as agenda setters and gatekeepers along the electoral cycle. Media attention tends to be more concentrated in a few issues during election time, especially because there are more stories about political parties and their electoral strategies. But newspapers do try to reinforce issue ownership of certain issues by the PP and PSOE during elections. *El País* is not focusing more on social issues or the environment during elections, as compared to routine times. The same occurs for *El Mundo*, which does not pay more attention to the economy even though one might expect it to from an issue-ownership perspective. We see that both newspapers reduce attention to a great number of policy issues during election times as they focus more attention on government operations and the electoral race itself.

Conclusions

By comparing the two most-read Spanish newspapers, we conclude that media attention is highly concentrated on a few issues. The attention is highly correlated among newspapers, suggesting that fears of "information silos" across different ideological groups in the electorate are unfounded. Newspapers inform the citizenry about what is going on in the political sphere, always focusing on the bad news and following different patterns. First, issue attention is always more concentrated on a few issues in the case of *El Mundo*, than for *El País*. This is because *El Mundo* pays more attention to crime and justice (e.g., domestic terrorism issues) and to political scandals. *El País* also pays a large share of attention to these issues but with much less intensity. Second, both newspapers attack their ideological opponents by always talking more about their opponents than their allies. The distinction is most striking when either prints stories on political scandals. For 2008 to 2011, *El Mundo* paid seven times more attention to the PSOE than the PP; while for the same period, *El País* paid twice as much attention to the PP than the PSOE. Newspapers give a one-sided view of the political debate, focusing more on their adversaries than their allies, always focusing on the faults, and reporting more about elite's differences than similarities.

Our results also show that both newspapers follow quite different paths when performing their tasks as agenda setters. The correspondence between issue attention in *El País* and *El Mundo* and parliamentary activities is high and significant when parties are in the opposition and for a large set of issues, especially those not related to the welfare state (with the exception of education). Coverage in *El País and El Mundo* is linked with the current activities of opposition parties, especially for issues related to the economy, unexpected and focusing events like food safety or environmental problems, and highly controversial issues like terrorism, education policy, and political decentralization. For the last decades, parties are using the media as a place for political confrontation, reinforcing the old political cleavages that have characterized Spanish politics for decades. Both the PP and PSOE follow the media only for highly controversial issues – like education in the case of the PSOE, or terrorism in the case of the PP – that may be rewarding in electoral terms. Finally, globalization and increasing market competition did not alter media coverage of politics nor the institutional features of the Spanish media system, which more than ever is characterized by political parallelism and the erosion of media freedom. In the next chapter we provide further information about the role of the media in issue attention by focusing on political decentralization as an issue.

Part III

Policy Issues in an Agenda-Dynamics Approach

6
The Politics of Decentralization

Political decentralization is one of the key issues in Spanish politics (Caminal and Requejo, 2010). From the transition to democracy to the present, an intense and highly controversial debate has raged about the territorial distribution of power, and this debate has consistently been at the core of differences among Spaniards both within and beyond the parliamentary arena. Spanish citizens and political elites have opposing ways of thinking about the limits of political decentralization. Further, the debate has become increasingly intense over time. At the end of the first decade of the new millennium, most Catalan citizens gave support to secessionist vindications, while the central government was increasingly talking about recentralization (Chaqués-Bonafont and Palau, 2011a).

For the first time, the symbolic dimension has come to dominate the debate – especially since Catalonia, while drafting its new Statutes of Autonomy (Estatuto de Autonomía) in the mid-2000s, formally defined itself as a nation. For most Catalan citizens this was simply a formalization of a reality that has been there for centuries. For others, it was a direct attack on the unity of Spain. Whereas the debate once centered on what policies would be decided at what level of government – with different regions of the country opting for different positions on this issue, and with great diversity across the various regions – increasingly today the debate has become dominated by highly symbolic issues such as what is a nation and who can fly which flag.

Extreme positions are not new – on average, fewer than 10 percent of Spanish citizens prefer a centralized Spain, and about 7 percent would prefer secession of their region from Spain, a percentage that increases up to 25 percent in the case of Basques and Catalans from the 1980s to the mid-2000s (Keating, 1998, 2001; Torcal and Medina, 2002; Muñoz,

2012). What is new is the growing support to both extremes by political elites and Spanish citizens in general (Guibernau, 2013). In less than ten years, the percentage of Spanish citizens that prefer a unitary state has almost tripled, while the percentage of Catalans in favor of secession has more than doubled (see Tables 6.1 and 6.2). These polarizing public attitudes have been accompanied by a shift in the nature of the public debate: Citizens' organizations pressing for independence have emerged as key actors in the policymaking process, transforming a policy process that was once dominated by political elites. In previous rounds of negotiation on these matters, debate centered on agreements between the governing party and the main opposition party in the 1980s; bilateral negotiations between the governing party and regional parties, which played pivotal roles in the formation of Spanish governments in the 1990s; or later, the complex negotiation between statewide and regional parties in the 2000s, including the Basque-focused *plan Ibarreche* (2004) and the Catalan Estatuto de Autonomía (2005).

Changing public attitudes and increasing social mobilization illustrate the failure of the so-called politics of consensus that traditionally dominated the debate about the relations between Madrid and the regions. Several trends document this general pattern. Parliamentary bills related to political decentralization become much more likely to fail as time goes on; political elites systematically use the Constitutional Court as a political venue to solve major disputes about this issue, delegating their responsibilities and promoting the judicialization of a "question d'Etat"; and parties intensify a confrontation strategy, using more and more state-regions relations as an electoral weapon. In sum, the question of regional autonomy has moved from one featuring accommodation to one of highly symbolic and politicized position-taking with extreme positions becoming more common in recent years.

Statewide parties strategically choose how and when to talk about political decentralization. In doing so, parties do not provide a coherent discourse or a clear set of ideas about how to deal with citizens' demands. Nor do they appear to search for the accommodation of different views and perspectives about what should be done about state-regions relations. On the contrary, in contrast to previous decades, the large, statewide parties adopt a strategy of attacking each other, focusing on major political scandals and conflicts as a means to maximize electoral rewards (Vallespín, 2012). And, when they are in the opposition, they accuse their rival governmental party of selling out the national interest in whatever compromises they might agree to with the regions. So state-regions relations have increasingly become just one in a number of areas

Table 6.1 Preferences toward the territorial organization of power, 2005

	Party vote in March 2004 election													
	Total	**PP**	**PSOE**	**IU/ICV**	**CiU**	**ERC**	**PNV**	**BNG**	**CC**	**EA**	**CHA**	**Na-Bai**	**Other parties**	**Below 18 in March 2004**
Central government without the regions	8.7	16.2	6.1	3.8	5.7				4.4		13		14.5	4.2
Estado de las Autonomías as it is now	51.4	65.4	50.4	36.7	15.2	1.0	10.3	39.2	33.3		21.7	3.1	24.6	56.8
More political autonomy for the regions	26.3	14.0	34.2	41.7	54.3	23.1	43.5	43.1	44.4	39.9	56.5	34.4	37	18
A state that recognizes the right of regions to become independent states	6.6	.7	3.7	14.9	22.9	75.0	43.5	15.7	4.4	60.1	4.3	62.5	16.7	12.4
Don't know	5.9	3.1	5.0	1.4	1.0	1.0	2.7	2.0	13.3		4.3		2.2	7.3
No answer	1.2	.6	.6	1.5	1.0								5	1.3
Total	100	100	100	100	100	100	100	100	100	100	100	100	100	100

Source: Centro de Investigaciones Sociológicas (www.cis.es) survey number 2610. Barómetro Autonómico December 2005. Organization of power question: "Here you have different alternatives about the territorial organization of the state. Please, which of these options respond to your preferences?" Partisan preference question: "Which political party did you vote for in the past general elections?" The survey covers all the territory of Spain. Respondents are citizens 18 years of age or older. The previous general elections were in March 2004.

Table 6.2 Preferences toward the territorial organization of power, 2013

	Total	PP	PSOE	IU/ ICV	UPyD	CiU	Other parties	Below 18 in November 2011
Central government without the regions	22.1	38.4	18.5	15.4	32.3	4	6.8	17.9
A state in which regions have less political autonomy than at present	12.2	15.0	10.2	13.3	21.5	2	8.6	16.1
Estado de las Autonomías as it is now	30.5	29.4	41.2	31.5	26.2	4	17.9	25.0
More political autonomy for the regions	13.1	6.1	16.0	23.1	6.2	32	22.2	16.1
A state that recognizes the right of regions to become independent states	8.5	1.3	3.3	8.4	1.5	54	34.0	12.5
Don't know	10.0	7.9	8.9	4.2	1.5	2	6.8	12.5
No answer	3.5	1.8	1.9	4.2	10.8	2	3.7	-
Total	100	100	100	100	100	100	100	100
(N)	2,472	554	481	143	65	50	162	56

Source: Centro de Investigaciones Sociológicas (www.cis.es) survey number 3005. Barómetro autonòmico November 2013. Organization of power question: "Here you have different alternatives about the territorial organization of the state. Please, which of these options respond to your preferences?" Partisan preference question: "Which political party did you vote for in the past general elections?" The survey covers all the territory of Spain. Respondents are citizens 18 years of age or older. The previous general elections were in November 2011.

on which political points can be scored between the government and the opposition. Our results clearly illustrate this strategy. Both the PP and PSOE more and more try to avoid the issue of state-regions relations while in government. Actually, Mariano Rajoy only devotes 2 percent of his first speech to deal with political decentralization, despite mass demonstrations and changing public opinion. In contrast, Felipe González devoted more than 23 percent of his first speech in 1982 to talk about this issue, Aznar more than 12 percent of his first speech in 1996, and Rodríguez Zapatero about 6 percent of his speech in 2004.

While in the opposition, political parties intensify their attention to state-regions relations. In fact, from 1982 to 2004, parties devoted an average of 7 percent of oral questions to this issue, while from 2004 this percentage increases to more than 14 percent, reaching 20 percent in some years. This increase is not explained by any changes by the regional parties, since parties like CiU or PNV always push this issue into the agenda no matter whether they are giving support to the government or not, and independently of whether an increasing number of Spanish citizens identify national-regional relations as one of the most important problems of the country. And the small parties are allocated only a small number of questions in any case. Rather, the increased attention to decentralization in parliamentary oral question sessions stems from a decision by the large statewide parties to try to embarrass the government and to force the governing party to take a position on this highly controversial issue, a topic that, while in government, that same party may well have attempted to avoid.

In all, we show in this chapter the intensification of the symbolic and economic dimension of the debate about political decentralization. One of the most controversial debates is focused on the identification of Catalonia as a nation, and the use of Catalan language, which generated one of the largest demonstrations ever in Catalonia in July 2010, immediately following the decision of the Spanish Constitutional Court about the *Estatuto* of Catalonia. At the same time, from the mid-2000s, the PP led by Mariano Rajoy and the Catalan governments have emphasized the economic problems associated with delegation of authority, leaving aside other concerns about policy transfers, which traditionally dominated the debate. The Spanish government has increasingly pointed to the regions as one of the main causes of the rising public deficit, arguing that the regions are to blame for a large part of the worsening economic crisis. For the Catalan government, there is a permanent injustice toward Catalonia in terms of public investment and fiscal transfers, eroding the region's capacity to fully exercise its political autonomy.

The radicalization of the political debate occurred at a time when Spain already had a comparatively high level of regionalism (Hooghe et al., 2008). The capacity of the regions to exercise authority in their territories has increased, especially after the fiscal reform of the late 1990s, but this has not softened demands for independence in the Basque Country and Catalonia. On the contrary, in the new millennium, we see how the go-along and get-along idea (e.g., the idea of *conllevarse* as explained by philosopher Jose Ortega y Gasset) that has characterized the relations between central government and the regions with strongest demands for autonomy for decades is replaced by a new relationship based on confrontation and bipolarization of the political debate (Gunther and Montero, 2009; Keating 1998, 2001). At the end of the chapter we argue the media has contributed to this transformation, always giving the bad news and increasingly emphasizing a negative tone in its coverage of state-regions relations.

The chapter is organized as follows. First, we provide a general overview of regionalization as a process before explaining the four waves of regulatory change. From here, we explain how and why the debate is always more focused on self-rule (the capacity of the regions to gain political authority in their territory) than shared rule (the capacity of the regions to be part of the policymaking process at the national level). Next, we provide new evidence about parliamentary activities related to political decentralization and test to what extent political parties use this issue for electoral purposes. In the final section, we provide new evidence about the role of the media in this process of increasing polarization and confrontation of the political debate.

A quasi-federal system of government

Over the past decades, Spain has acquired almost all the features attributed to a federal state. Policy responsibilities are increasingly fragmented, shared, or/and divided across levels of government as a result of a long and controversial process of negotiation between the Spanish and the regional authorities. The Spanish Constitution of 1978 defines a quite open and flexible system of distribution of political autonomy across levels of government – the so-called *Estado de las Autonomías* – designed to accommodate the historical demands of some of the territories for more power with the vindications of those who support a single national system for all. This last position, that of the unity of Spain, is defended by a vast set of actors – especially the military, hardliners of the Franco regime, and also most statewide parties. All these negotiations, of course, have taken place during a period of transition to democracy, which gave significant legitimacy to those who argued that centralized power had been misused in the past (Rodríguez-Aguilera and Morata, 2008). The flexibility of the "state of autonomies" is observed on the fact the Spanish Constitution allows for several different routes for a region to become autonomous: The slow track, defined by Article 143; the fast track, defined in Article 151; and the special track defined in the second transitional provision. The Constitution also sets out the possibility of different levels of issue jurisdiction.

In 1985, the Spanish territory was divided into 17 regions despite the fact that, when drafting the Constitution, observers expected that only the three so-called historical regions (the Basque Country, Catalonia, and Galicia), which had already enjoyed a high level of political autonomy before Franco's dictatorship, would initiate the process to become "autonomous communities." In less than a decade, Spain adopted a hybrid system of distribution of authority across territories, radically transforming the unitary model that had characterized the Franco dictatorship for more than four decades. In fact, Spain exhibited among the highest levels of political decentralization, particularly in terms of self-rule (that is, the authority exercised by regional governments over local residents) among all advanced democracies (see Hooghe et al., 2008). This followed from the fact that, even if only three historically autonomous regions had strong claims to local rule, all 17 regions adopted certain institutional arrangements (see Aja, 2014). After 1985, all regions were empowered with a general purpose administration independent of any veto by the Spanish government; a directly elected regional Parliament, which appoints the president of the region; and authority over a wide range of policy issues (though in this case there are some important differences across the regions; see Table 6.1).

The seven regions that followed the special or fast route – Andalucía, the Canary Islands, Catalonia, Galicia, Navarra, the Basque Country, and Valencia – also reached a (provisionally) higher level of issue jurisdiction, especially in relation to social welfare issues and, with the exception of the Canaries, Navarra, and Valencia, they have the unilateral capacity to define the timing of legislative elections, while all others must call for elections on the same day (the last Sunday of May every four years). Asymmetries also exist in terms of official languages (the Spanish Constitution recognizes Catalan, Galician, and Basque as well as Castillian), the legal system (Catalonia has a separate civil code), police (the Basque Country, Navarra, and Catalonia have their own regional police), and fiscal autonomy (the regional government's capacity to determine its revenues unilaterally; Aja, 2014).The Basque Country and Navarra collect major taxes in their territory and pay a quota to the central government (called *concierto*) for the performance of national-level government functions. That is, in those regions, the Spanish government does not even collect taxes. Clearly, decentralization in Spain is significant but also quite uneven as some regions have high levels of autonomy whereas others have less. Further, the variation is not only geographic but by policy domain as well; some policies are more centralized; others, largely attributed to the regional authorities.

While some Spanish regions exert significant power over those living in their areas, the ability of a regional authority to affect national politics is very low compared to other federalized systems. For example, the Spanish Senate, which includes both directly elected members as well as those appointed by the regional authorities, has a very marginal position in the legislative process. Regional governments have no formal authority to affect national legislation or constitutional changes, and most of the decisions taken by the multilateral committees formed by the central and regional governments have no binding authority. Thus, we see a high degree of "self-rule" within the regions but very little "shared rule" at the national level. This is, in part, because of institutional rules in the Spanish constitution, but more generally because of the preferences of national-level political actors who make significant efforts to promote (or prevent) transfers of issue jurisdiction (e.g., self-rule), but avoid promoting any policy change that would lead toward more shared rule, as we explain in the rest of the chapter.

Four waves of policy change

The construction of Spain's peculiar quasi-federal system of governance reflects four waves of policy change. The first starts in 1979, with the new Spanish Constitution and the Estatutos de Autonomía of the Basque Country and Catalonia, and ends in 1983, with the completion of regional elections in all 17 regions. In this period, the two large, statewide political parties – the UCD (governing party) and the PSOE (the main party in the opposition) – made clear their desire to enact limits on the *Estado Autonómico*. Both parties unsuccessfully opposed the extension of political decentralization to other regions beyond Catalonia and the Basque Country, and once this was inevitable, they attempted to limit political authority by passing the Organic Law for the Harmonization of the Autonomy Process (*Ley Orgánica de Armonización del Proceso Autonómico*, LOAPA) in 1982, a few months before the general elections. In 1983, most of the content of the LOAPA was judged unconstitutional by the Constitutional Court, one of the key political factors in the consolidation of the *Estado Autonómico*.[1] So from the start we see important differences depending on the region, and a general refusal by the main statewide parties to accede to decentralization demands, especially from those regions not historically independent from Madrid.

The second wave starts in the late 1980s. Following the process laid out in Article 143 of the Spanish Constitution, five years after their founding as regional governments, the ten regions having taken the slower track toward increased autonomy had the right to increase their political powers according to the limits imposed by Article 148 (which lists the policy domains that the regions may assume under their jurisdiction) and Article 149 (which lays out the issues to remain under the exclusive jurisdiction of the Spanish state). After more than two years of intense debates, the PSOE and PP arrived at an agreement by which these ten regions increased their powers to almost the same level of the seven regions that had taken the faster track to greater local control. This agreement was formalized with the Organic Law on Regional Transfers (*Ley Orgánica de Transferencias, 9/1992*) in 1992, and various reforms to the Statutes of Regional Autonomy (Estatutos de Autonomía), all passed as organic laws before 1995. Regulatory changes which followed these laws later completed the homogenization of issue jurisdictions across all territories, including welfare-related issues such as health and undergraduate education, which were formally transferred to those regions not already controlling them in 2000 (Subirats and Gallego, 2002). As a result, by 2000, the only differences in terms of issue jurisdiction were related to fiscal autonomy (which has always been greater in the Basque Country and Navarra), language, and the civil code as described above (Aja, 2012; Caminal and Requejo, 2010).[2]

The process of homogenization defined in the Decentralization Agreements (*Pactos Autonómicos*) of 1992 was completely at odds with the preferences of the historical nations of Catalonia, the Basque Country, and Galicia. In fact, rather than putting an end to the debate about political decentralization, the *Pactos Autonómicos* are the point of inflection after which decentralization became increasingly polarized. After this point, the so-called historical regions increasingly felt the need to advocate for an asymmetric model of political decentralization, whereas the other regions, always governed either by the PP or PSOE, increasingly defended a symmetrical (or one-size-fits-all) federal model. The historically recognized regions of Catalonia, Galicia, and the Basque Country saw homogenization as a strategy of the central government to limit any future expansions of issue jurisdiction, fiscal autonomy, and capacity to participate in the EU policymaking process, among other things. This is because only the historical regions show any interest in these matters. If decentralization is to be on an equal footing, those regions with great demand and support for more regions cannot go faster than others. Of course, for those who support a stronger national government, this is exactly the point: to use the concept of homogenized decentralization to limit the scope of the maximum decentralization that any single region might achieve.

As a result of these developments, political decentralization consistently occupied a predominant place in the governmental agenda of José María Aznar (1996–2004). The PP government put forward a regional fiscal reform, according to which regions have the capacity not only to collect taxes, but also the ability to regulate and set the bases for major taxes including the income tax. This tax reform was first implemented in 1996, just a few months after winning elections, as a result of the agreement (*Pacto del Majestic*) between the PP (governing at that point as a minority government) and the Catalan-based CiU party, on which the government relied to maintain its parliamentary majority.[3]

The influence of regional political parties in the national government depends not so much on their overall level of electoral support (after all, it is insignificant outside of their regions, so it cannot be a large part of the national total), but rather on their capacity to impose some of their preferences when they are pivotal in the formation or support for the government at the national level (Chaqués-Bonafont and Palau, 2011a). This was the case from 1993 to 1996, when Felipe González (PSOE) was governing under minority with the support of the Catalan conservative nationalist party (CiU); from 1996 to 2000, when José María Aznar governed under minority with the support of CiU and also the PNV; and also from 2004, when José Luis Rodríguez Zapatero (PSOE) governed with the support of the Catalan nationalist party of the left (ERC) and the statewide far-left party Izquierda Unida (IU). In fact, the 1996 fiscal reforms were powerful indicators of the power of regional parties when the main statewide parties lack a unilateral majority in Parliament (Keating and Wilson, 2009). These fiscal reforms, recognized by Hooghe et al. (2008) as leading to important increases in the decentralized character of Spanish government, were in fact already negotiated with the previous government of Felipe González, and when the PP won elections in 1996 the regional parties insisted on going through with the fiscal decentralization plan (Bosch et al., 2010). Thus, a few votes moving from PSOE to PP led to a change of government, but did not dilute the power of the regional parties at that time, and they forced the minority government of the right to enact a reform similar to that which they had negotiated but not enacted with the previous minority government of the left.

These examples show the occasional importance of the regions at the national level. Note, however, that this influence depends on the minority status of the government and its reliance on the parliamentary votes of regional parties present in Madrid. In contrast to how regional authorities are often represented at the national level in other federal systems, the Spanish Constitution offers almost no institutional role for shared power of the regions themselves; the influence is through the party system and the composition of the government. With powerful parties holding exclusive control of a parliamentary majority, regional influence would be very weak, as neither of the major statewide parties endorses these reforms voluntarily. So, important reforms giving greater powers to the regions were implemented not because the statewide parties preferred to do so, but because their control of a parliamentary majority depended on the support from Basque or Catalan parties, which extracted this as the price of their support.

Finally, the fourth wave of decentralization begins in 2001 when the Basque regional government introduced the so-called *plan Ibarreche*, proposing revisions to the Statutes of Autonomy; while these proposals were hotly debated in Madrid, they were eventually shelved as the statewide parties refused to accept them. This opened a new wave of politicization, characterized by an intense and polarized debate about the relationship between the historical regions and the rest of Spain. The epicenter of the debate occurred during the first legislature of Rodríguez Zapatero, governing under minority with the support of the Catalan left (ERC) and the Far Left (IU). The legislature starts with the debate of the *plan Ibarreche* in 2004, continues with the passing of the Catalan Estatuto de Autonomía as an organic law in the Spanish Parliament, and its approval in referendum in Catalonia in June 2006, and ends with one of the most polarized general election campaigns in 2008, in which the *problema autonómico* keeps being one of the key issues in the political agenda.

The radicalization and polarization of the debate about the *Estado Autonómico* transcended the parliamentary arena, and became an important object of mass mobilizations. The setoff was the decision of the Constitutional Court that declared the unconstitutionality of some parts of the Estatuto de Autonomía of Catalonia, after four years of intense and controversial debates. More than a million people gathered in the streets of Barcelona on July 10, 2010, with the slogan "*Som una Nació, Nosaltres decidim*" ("We Are a Nation; We Can Decide"). An issue that mobilized less than 50 percent of Catalan citizens for the referendum of the Estatuto de Autonomía on June 18, 2006, was gathering more than a million people in the streets of Barcelona four years later. These large demonstrations and popular consultations at the local level were led by civil society organizations like the *Plataforma pel Dret a Decidir, Òmnium Cultural*, and the *Assamblea Nacional Catalana*. For the last three years on every September 11, the national day of Catalonia, more than a million people have gathered in the street to vindicate their right to become a new member state of the EU, and more than half of all Catalan municipalities organized popular consultations about people's willingness to became politically independent from Spain beginning in September 2009.

Changing public attitudes are highlighted in Figure 6.1, which compares results of the three regional barometers (*Barómetro Autonómico*) developed in 2005, 2010, and 2012, and the survey about institutions and regions (*Institutiones y Autonomías*) developed in 1998.[4] In 2012, the percentage of citizens of Catalonia identify themselves as only Catalans, or more Catalan than Spanish, reaching more than 50 percent, matching levels seen elsewhere only in the Basque Country. This constitutes a radical change from the late 1990s when 35 percent of Catalan citizens identify themselves as only or more Catalans. In other regions, the percentage of citizens that identify themselves as only or more of their region remains more or less the same over time, with some variations across regional governments. That is, for most of the regions, identity shifts do not occur, with the exception of Catalonia.

However, as Figure 6.2 illustrates,[5] Spanish citizens increasingly show their disagreement with the existing quasi-federal model that has characterized Spain for decades. In October 2007, almost 60 percent of Spaniards preferred the existing *Estado autonómico* to any other form of territorial distribution of power, while in February 2014, this percentage was less than 35 percent. In contrast, the percentage of citizens that favor recentralization, (i.e., less autonomy) for the regions, increases from 8.7 percent in 2007 to more than 21 percent in 2014. This trend is more or less the same for all regions but the Basque Country. In the Basque Country, for example, fewer than 3 percent prefer a unitary state, and this number does not vary over time. In Catalonia, as Figure 6.3 illustrates, in 2012, the percentage of citizens that supports the right of regions to become an independent state is more than 40 percent, which represents a 75 percent increase in less than two years. So we see very important regional differences with a growth of independence demands over time.

These changing trends are also clearly illustrated in Tables 6.1 and 6.2, which provide further information about citizens' preferences about territorial distribution of power controlling by party preferences (measured by voting preferences in the last general election). In 2013, 38.4 percent of PP voters, 18.5 percent of PSOE voters, and 15.4 percent

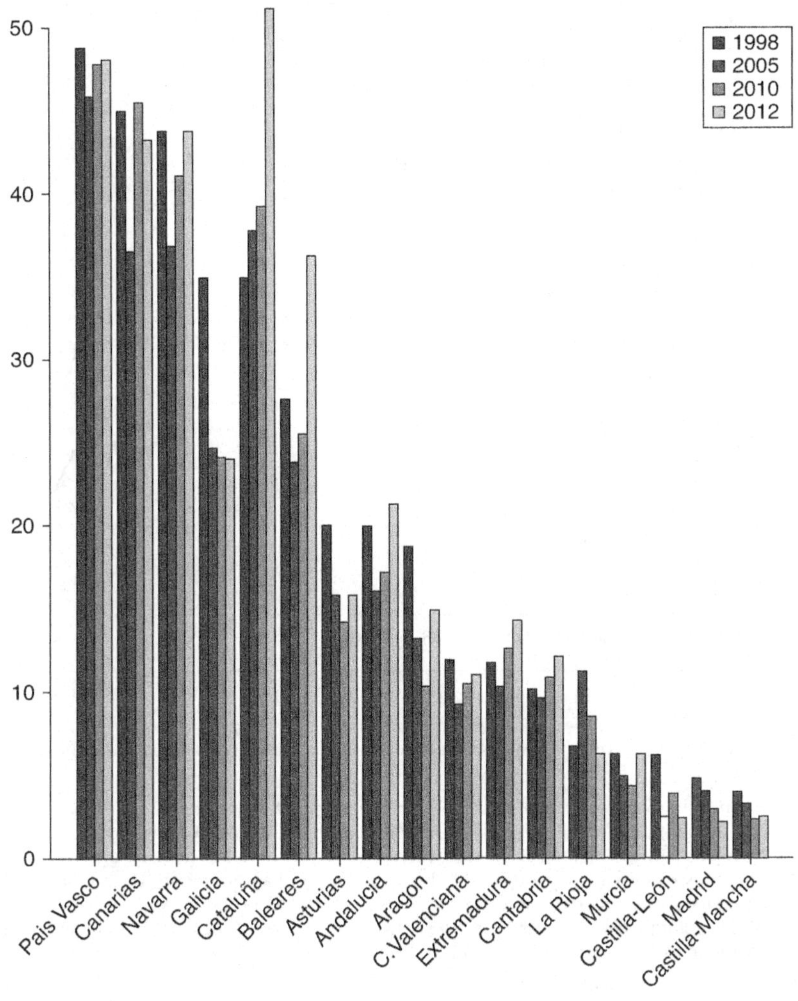

Figure 6.1 Regional identity, over time and by region

Note: For each region, bars indicate the percentage of citizens that identify themselves as more or only from their region, as opposed to from Spain.

Source: *Centro de Investigaciones Sociológicas* (www.cis.es): survey number 2286, Instituciones y Autonomías (1998), survey number 2610, Barómetro Autonómico I (2005), survey number 2829, Barómetro Autonómico II (2010), survey number 2956, Barométro Autonómico III (2012).

of IU voters prefer a unitary model. Compared to 2005, these numbers represent more than double the numbers for the PP, triple for the PSOE, and five times more in the case of IU. At the other end of the spectrum, the percentage of CiU voters in favor of secession increases up to 54 percent in 2013, double the level of 2004.

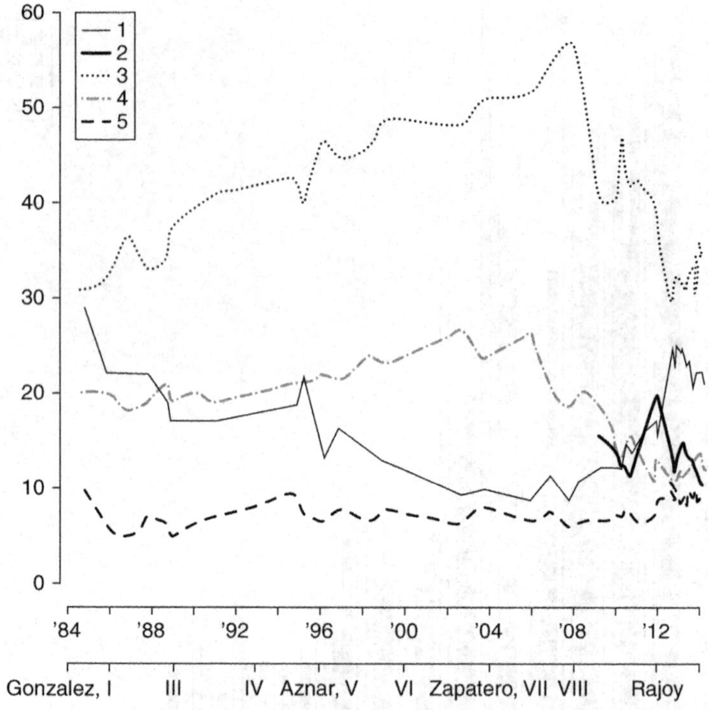

Figure 6.2 Preferences of Spanish citizens about the territorial organization of power

Note: Lines show the percent of citizens expressing support for each of the following allocations of political power between the central government and the regions: 1. Central Government without regions; 2. A state where the regions have less political autonomy; 3. Autonomous regions as they are now; 4. Increased political autonomy for the regions; 5. A state that recognizes the right of regions to become independent (option only available after 2008). The figure shows a dramatic decline in those thinking that the current arrangement is best, with an increase in polarization after 2009.

Source: Centro de Investigaciones Sociológicas (www.cis.es).

In short, now more than ever, political decentralization is one of the most controversial issues in the Spanish political agenda. For the first time since the transition to democracy, the government in Madrid is talking about recentralization, while Catalans and the Basques are vindicating their right to become new member states of the EU. The central government argues the regions badly managed their public accounts, wasting public resources, and directly contributing to aggravating the economic crisis – regions' total debt accounts for more than 140 billion euros in

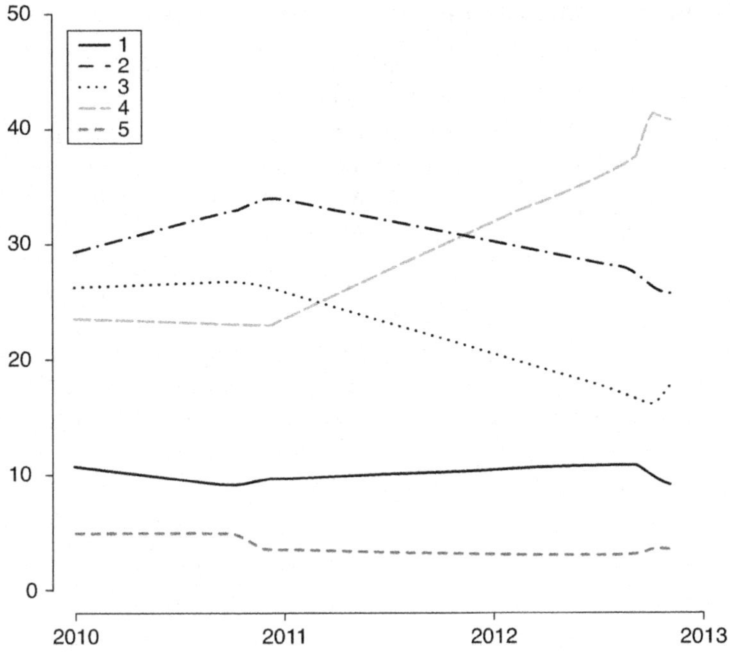

Figure 6.3 Preferences of Catalan citizens about the territorial organization of power, 2010–2012

Note: As in the previous figure, lines show the percent of citizens of Catalonia expressing support for each of the following allocations of political power between the central government and the regions: 1. Central Government without regions; 2. A state where the regions have less political autonomy; 3. Autonomous regions as they are now; 4. Increased political autonomy for the regions; 5. A state that recognizes the right of regions to become independent. Support for the status quo declines after 2011 as option 5, independence, becomes more popular.

Source: Centro de Investigaciones Sociológicas (www.cis.es).

2013. In contrast, regions like Catalonia blame the Spanish government not only of mismanagement – especially related to transportation policy, like high-speed train infrastructures, or unemployment – but especially on the long-standing inequalities generated by the redistribution of tax revenues from rich to poorer regions. As a result, traditional regional demands for more fiscal sovereignty – similar to the *conciertos* in the Basque Country and Navarra – have been replaced by secessionist vindications, dominating most of the political debate. This review of four stages of policy change clearly points to an increasingly polarized country with less and less common ground across the political spectrum.

Self-rule and shared rule

Attention to regionalization always occupies an important share of the attention in and out of the parliamentary arena. Figure 6.4 summarizes issue attention to regionalization from 1982 to 2011 in three different venues: speeches, oral questions, and executive bills. The figure provides information not only about the debates dealing with the distribution of power between the central government and the regions (subtopic 2070 in our databases) but also on all parliamentary activities that focus on regional concerns (what we called the regional political agenda identified with a dummy variable across all subtopics; see the Appendix). This includes oral questions about whether the Spanish government is willing to transfer unemployment management to the regions, or to promote the use of other co-official languages in public administration;

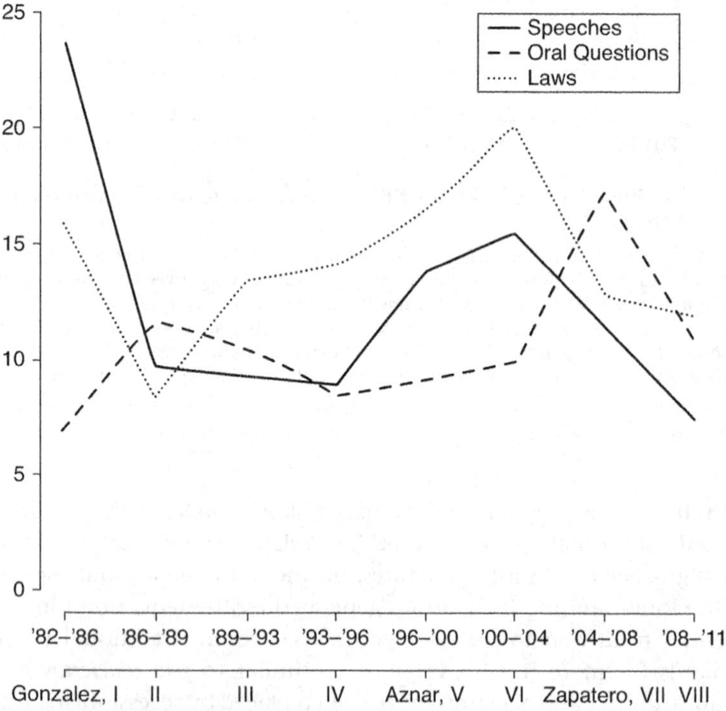

Figure 6.4 Attention to regionalization

Note: The figure shows the percentage of speeches, oral questions, and laws devoted to the issue of regionalization, by year.

Source: Spanish Policy Agendas Databases (www.ub.edu/spanishpolicyagendas).

the introduction of bills designed to modify the structure, goals, and composition of the Constitutional Court; or any mention in the annual speech by the Prime Minister about fiscal policy relating to the regions; or the reform of the Senate as a chamber that truly represents the different territories, to give some examples.

First, attention is especially large during the first wave of regionalization and particularly along the first legislature of González. More than 23 percent of his speeches and 15 percent of the executive bills from 1982 to 1986 were devoted to political decentralization, decreasing to an average 9 percent and 12 percent in the case of speeches and executive bills, respectively, in the following years. Once Aznar won election, attention to regionalization increased again up to 15 percent in the case of speeches and 20 percent for executive bills. Aznar spoke and took decisions about the regional fiscal reform and the implementation of the *café para todos* agreement in the late 1990s.

Interestingly, Rodríguez Zapatero was the head of the executive who devoted the least attention to regionalization, less than 9 percent of the speeches and 11 percent of executive bills. In contrast to previous Prime Ministers, he ignored the dimension of issue-transfers and fiscal autonomy that had dominated the debate in the 1980s and the 1990s, respectively (see Figures 6.5 and 6.6). His few references to the regions mainly referred to the need to foster cooperation between them and the central government, the reform of the Senate, or the importance of some regions for the country's governability. In practical terms, however, these proposals were simply rhetorical and were never implemented. Rodríguez Zapatero fostered this strategy of non-decision in a political context characterized by increasing confrontation between one of the regions from which the PSOE gains significant electoral support, Catalonia, and most of the rest of Spain. In fact, the reform of the Estatuto de Autonomía of Catalonia was led by Pasqual Maragall, the first socialist to become president of Catalonia, after more than two decades of CiU government led by Jordi Pujol.

The information provided so far illustrates that in Spain, as in most other advanced democracies, regionalization is mainly about increasing regional self-rule, more than shared rule. This is explained by several factors. On the one hand, institutional friction associated with changes in self-rule is lower than for changes in shared rule (Hooghe et al., 2008). Delegation of issue jurisdiction always follows a complex and controversial negotiation between statewide parties and regional forces. Still, transfers of issue jurisdiction or even fiscal reforms can be put into practice by passing a law, which may involve the party in government and another statewide party (either the main party in the opposition, as in the 1980s

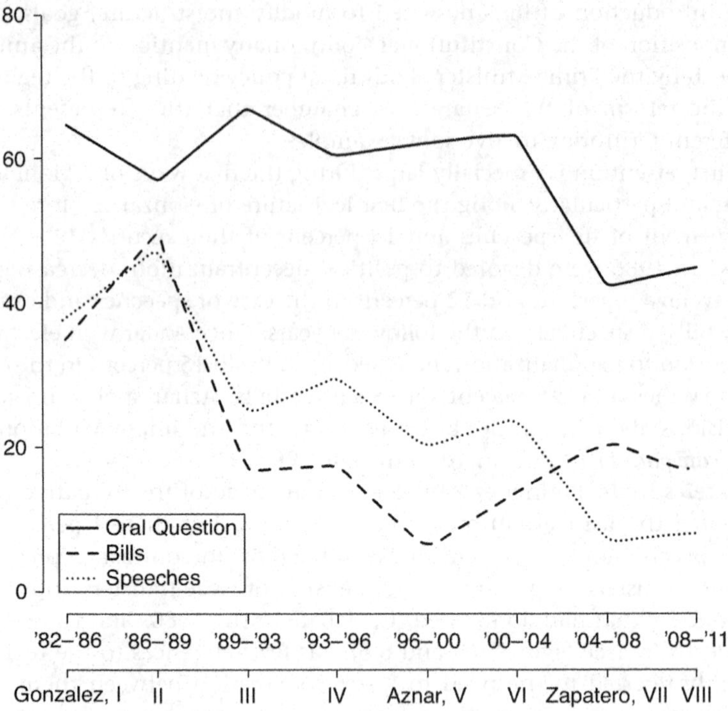

Figure 6.5 Attention to fiscal autonomy as a percentage of all attention to regionalization

Note: The figure shows the percentage of all oral questions, bills, and speeches focusing on fiscal autonomy of the regions (e.g., taxing authority) as a percentage of all of those that deal with regional issues.

Source: Spanish Policy Agendas Databases (www.ub.edu/spanishpolicyagendas).

and early 1990s with the *Pactos Autonómicos*) or a regional political party (like the CiU in the late 1990s). On the other hand, to formally institutionalize the participation of the regions in the Constitutional Court of Justice or to empower the second chamber as a venue of political representation for the regions, would require, among other things, amendments to the Constitution and thus the agreement of a larger set of policy actors, with quite different visions about the desirability of powerful regional voices within the national government.

However, the predominance of self-rule over shared rule cannot be explained in full by institutional factors; it also involves policy preferences. Shared rule is not a goal in itself either for the statewide or the

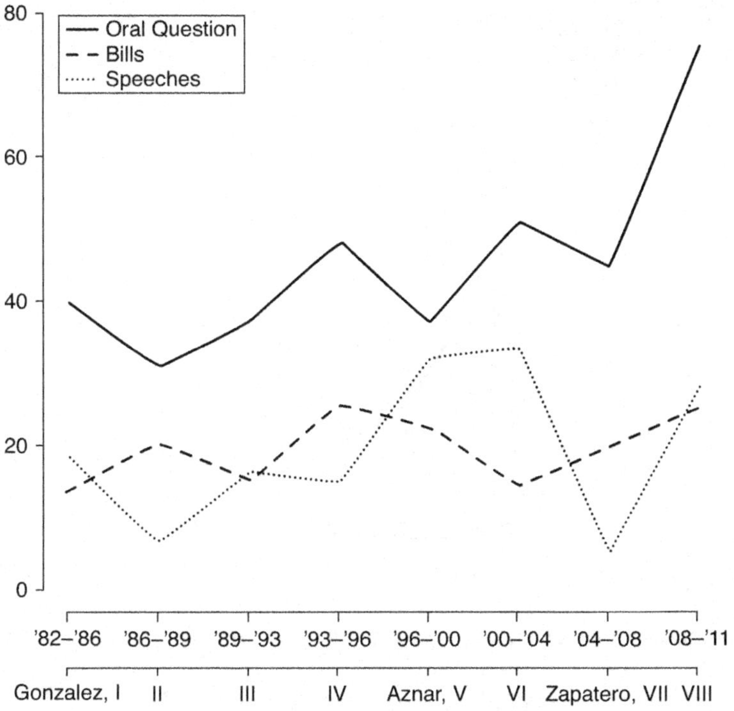

Figure 6.6 Attention to issue transfer as a percentage of all attention to regionalization

Note: The figure shows the percentage of oral questions, bills, and speeches focusing on issue transfers (e.g., which policy domains are in the competency of the national v. regional governments) as a percentage of all of those that deal with regional issues.

Source: Spanish Policy Agendas Databases (www.ub.edu/spanishpolicyagendas).

regional political parties. Neither Catalan nor Basque parties actively promote the consolidation of a set of federal institutions through which the 17 regions – in cooperation with the central government – define, coordinate, and implement policy goals. Actually, as Figure 6.7 illustrates, only 20 percent of executive bills are about shared rule, and most of them are registered during the 1990s by regional parties or the far left (the PP and PSOE do not introduce parliamentary bills about shared rule) to deal with internal issues related to the functioning of public administration, such as the roles of regional governors (*gobernadores civiles*) and the Constitutional Court.

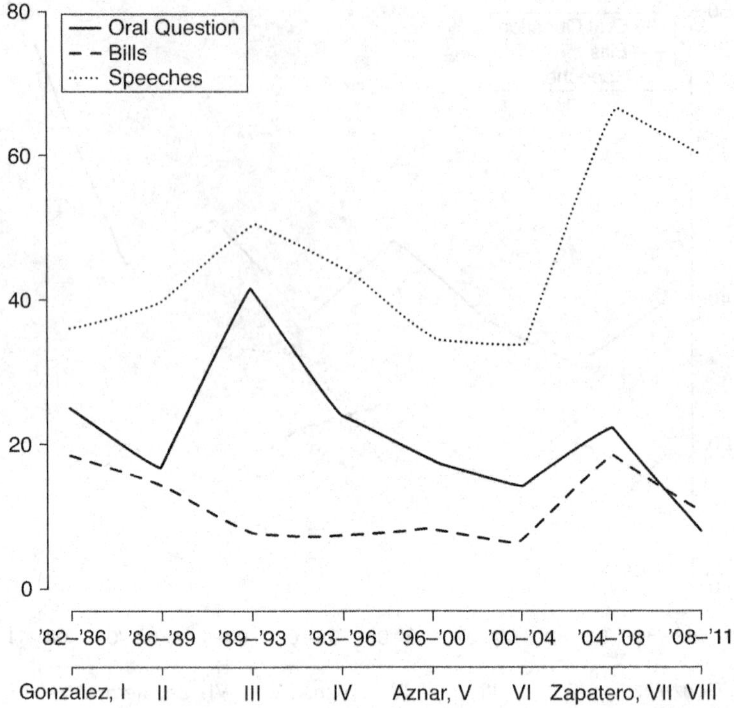

Figure 6.7 Attention to shared rule as a percentage of all attention to regionalization

Note: The figure shows the percentage of oral questions, bills, and speeches focusing on shared rule as a percentage of all of those that deal with regional issues.

Source: Spanish Policy Agendas Databases (www.ub.edu/spanishpolicyagendas).

Regional parties also ask most of the oral questions about shared rule, but in contrast to the case of bills, most of them refer to EU affairs. Almost 60 percent of all the oral questions asked by regional parties related to shared rule are linked to the EU (whereas only 10 percent and 4 percent of such questions coming from the PP and PSOE, respectively, have a connection to the EU). From the earliest days of Spanish entry into the EEC in 1986, regional parties, mainly CiU, CC, and all those in the *grupo mixto* have consistently and strongly argued for the right of regions to directly participate in EU decision-making processes.

For regional governments, the EU generates a dual democratic deficit. First, the EU fosters a process of recentralization in which regional governments lose their capacity to take decisions upon issues like agriculture, the environment, and/or industrial policy, which are now in the hands of the EU. Second, this process of recentralization is not accompanied by an increasing access of the regions in the EU policymaking process. On the contrary, regions find it increasingly difficult to control the Spanish government on EU affairs, and to make it act according to their preferences (Moravsick, 1994). As a result, since 1986, regional governments initiate a strategy to reduce this dual democratic deficit. One of the first steps is the creation of the *Conferencia para asuntos relacionados con la Comunidad Europea* (Conference on Issues Concerning Relations with the European Community, or CARCE, later called CARUE) through which the central government, led by the Ministry of Foreign Affairs, and the 17 regions deal with EU matters in general (Mateo and Morata, 2007; Arregui, 2009).[6] Later, in 2004, the regions participated in several committees of the European Commission, and from 2004, regional governments became part of the working groups of the Council of the EU (four representatives in the areas of education, culture, environment, competiveness, labor, social policy, and gambling from 2006).[7]

These reforms did not resolve the lack of access of regional governments into the EU policymaking process, among other reasons because only one or two members represent all regions in the EU meetings, though the 17 regions have different preferences toward EU-related policy issues. Further, not all regions are eager to participate intensively at the EU level. To participate in a committee of the European Commission or any other formal venue at the EU is only seen as an opportunity to foster policy change according to regional interest by some regions. This is especially the case when the central government is not responding to the demands of some regions, as was argued in the examples of the "Mediterranean corridor" (high-speed train connecting Spain with the rest of the EU by the Mediterranean Coast) or the national hydroelectric plan (see Mateo and Morata, 2007).

In short, most decisions taken about political decentralization are designed to foster self-rule, while there is little shared rule at the national level. While in office, Prime Minister Rodríguez Zapatero highlights in some of his speeches the need to transform the institutional features of the Spanish democracy into something more similar to a federal state, with a Senate that fully develops a role of representation of the territories. But these ideas are never put forward and have little support in the PP nor among the regional parties, whose main concern is to foster their participation in the EU policymaking process. So when we discuss decentralization or federalism in Spanish politics, it is largely about the power of regions to rule their citizens in certain policy domains within their territories, not about the influence or power of local communities and regions in the national government.

Politicizing political decentralization

Statewide parties always devote less attention to political decentralization than regional parties. Less than 7 percent of the oral questions asked by the PP, PSOE, or IU deal with issues related to regions, while CiU, PNV, CC, and ERC commit between 35 and 41 percent of their questioning time to these issues – percentages are quite similar in party manifestos before general elections as well. Statewide parties, do not devote much time when referring to specific issues like foreign trade or defense to regional concerns. On the contrary, regional parties tend to link most oral questions to regional issues.

One way to illustrate this is to measure differences in the mean percentage of questions with and without regional content by political party. As summarized in Table 6.3, mean differences are always significant for the case of statewide parties without many differences across issues. That is, there is no single issue for which the percentage of oral questions with regional content is significantly larger than oral questions without regional content. Note that these results do not only apply for large parties like the PP and PSOE but also small parties like IU. In contrast, regional parties significantly ask more oral questions referring to regional concerns and this is especially the case for the PNV. The percentage of oral questions asked by the PNV about civil rights, agriculture, crime, commerce, defense, and government are significantly linked to regional concerns, while in the case of CiU, this result applies for the case of transportation, education, social policy, R&D, and public lands. That is to say that for the statewide parties, topics are not regionalized; for the regional parties, almost all policy issues have a regional character.

We arrive at similar conclusions when comparing the fragmentation of the regional political agenda; that is, to what extent regional concerns are diversified across a vast set of issues. Fragmentation is larger for the case of regional parties than the rest (not shown here). No matter whether an issue has shared jurisdiction or not, parties like CiU and PNV would ask questions that in most cases relate to regional concerns. On the contrary, fragmentation of the regional agenda of statewide parties is quite low, indicating that most of the attention to regional concerns is concentrated on fewer issues. That is, the one question that CiU or PNV have the right to ask in a parliamentary session is systematically linked to regional concerns, no matter what the substantive content of the question. No matter if discussing health, transportation, or international affairs, these parties most often make a link to their region. This is an important difference in relation to the rest of the parliamentary

Table 6.3 Comparing attention to different topics with and without regional content

	Mean differences: regional content – nonregional content						
Topic	**PP***	**PSOE***	**IU***	**CiU**	**Sig.**	**PNV**	**Sig.**
Economy	−80.24	−52.04	−61.97	−43.30	.000	−6.90	.470
Rights	−92.05	−93.45	−68.13	−34.49	.000	−32.18	.000
Health	−85.17	−85.24	−34.61	−28.16	.010	−10.35	.290
Agriculture	−88.86	−83.58	−41.03	−19.47	.080	−31.72	.000
Labor	−89.02	−79.07	−92.31	−74.71	.000	−20.69	.090
Education	−87.14	−82.52	−53.85	−4.94	.630	18.47	.090
Environment	−81.36	−82.02	−32.69	18.39	.020	−3.45	.700
Energy	−92.82	−83.48	−67.69	−18.39	.060	2.30	.780
Transportation	−79.76	−88.70	−54.36	51.7	.000	2.53	.830
Crime & justice	−93.26	−85.28	−62.56	−31.77	.010	−33.09	.000
Social policy	−84.58	−77.36	−42.31	−56.33	.000	−6.89	.310
Housing	−67.82	−71.09	−38.46	−9.19	.330	−3.45	.410
Commerce	−94.06	−79.50	−44.23	−41.77	.000	−27.59	.010
Defense	−98.02	−94.43	−80.77	−30.46	.010	−25.29	.010
Research & development	−91.04	−78.63	−34.61	−22.42	.050	10.35	.320
Foreign trade	−75.86	−55.17	−11.54	−17.24	.023	−3.45	.326
International affairs	−93.80	−91.52	−76.92	−27.59	.010	−16.75	.150
Government	−74.98	−76.34	−63.94	−21.42	.060	21.87	.010
Public lands	−72.26	−58.27	−26.93	8.05	.460	10.35	.240

Note: Entries refer to the difference in the percentage of oral questions by deputies of the party indicated with regional content minus those without regional content. Negative values mean that most questions are devoid of regional content. *For the PP, PSOE, and IU, each difference is statistically significant at the level of .000. Data cover the period of 1982 through 2011.

Source: Spanish Policy Agendas Databases (www.ub.edu/spanishpolicyagendas).

agenda, where fragmentation is always larger in the case of statewide parties than regional parties (see Chapter 4).

Finally, as one might expect, regional parties show special concern for issues that affect them specifically, like immigration – Coalición Canaria asks almost 30 percent of all the questions related to rights, which mostly deal with the problems generated by illegal immigration. Similarly in the area of transportation policy, most questions are asked by CiU in relation to all types of infrastructure, but mainly high-speed trains because of their concern with the high-speed train linking Catalonia with the rest of Europe. In crime and justice, the PNV asks the same number of questions as statewide parties, mainly related to internal security issues affecting the Basque area.

In short, attention to "regional concerns" and political decentralization differs substantially among political parties and these differences are not linked to issue jurisdiction. Political parties do not only pay attention to those issues whose jurisdiction has been transferred to the regions, but also they link regional questions to a vast array of issues, from education, to transportation or even foreign policy. In doing so, the parties pursue very different goals. Regional parties have issue ownership about this issue, and despite their ideological differences (in terms of left-right cleavages), they share a common understanding about state-regional relations.[8] The electorate identifies them as the defenders of regional interests. Any achievement regarding transfers of issue jurisdiction or fiscal reform is attributed to their ability to negotiate vis-à-vis their opponents in the Spanish Parliament.

At the same time, statewide parties try to avoid regional issues while they are governing and only raise attention in the parliamentary arena while they are in the opposition. They do this not because they have become supporters of decentralization, but as a means to force the governing party to take a position about highly controversial issues, such as the identification of a region as a nation or the limits of the transfers of issue jurisdiction – topics about which the electorate is greatly divided. As we argued in Chapter 4, joined with terrorism, the question about the unity of Spain is one of the issues for which political parties replace the strategy of issue ownership by one of embarrassing the government, forcing it to discuss policy failures or politically unpopular issues.

Our goal here is to go further into the analysis presented in Chapter 4 in which we compared issue attention between the PP and the PSOE when they are governing or in the opposition. To do that, we developed an OLS regression model in which the dependent variable is the percentage of oral questions asked in a given month by a particular political party that has regional content and the independent variables are: (1) the percentage of citizens that identify nationalism as the Most Important Problem of the country, lagged one month; (2) the institutional position of a political party in the government-opposition game (governing or giving support to the government = 1, in opposition = 0); (3) agenda capacity, measured by the number of oral questions asked by that political party in the same month; and (4) a dummy variable = 0 before 2008 and = 1 for 2008 and beyond. Recall from Chapter 4 that from 1996 to 2008 the number of questions asked by the governing and the main party in the opposition is the same. From 2008, the Congreso de los Diputados changed the rules regarding an oral question, which implies among other things a decline in the number of questions asked by the governing party. The introduction of this variable controls for this variation. The period we take into account is from 2001 to 2013, because MIP data is only fully available during this period.

Table 6.4 summarizes results. First, attention to political decentralization by regional parties is unrelated to changes in public opinion. The percentage of oral questions on regional issues posed by the CiU does not increase with changes in public opinion, mainly because they are responding to their electorate (which in most cases considers nationalism) and to secession not as a problem but a goal in itself. In contrast to this consistent attention by regional parties, statewide political parties increase their attention to political decentralization when they are in the opposition, and especially, when an important part of the electorate identifies nationalism as one of the most important problems of Spain. The percentage of oral questions asked by the PSOE decreases by 2.47 when it is governing, with the equivalent number for the PP being 2.7.

Table 6.4 Explaining the percentage of oral questions with regional content

	B	Sig.	R^2
PP			.533
MIP	3.878	.001	
Governing	–9.380	.001	
Agenda capacity	.794	.000	
PSOE			.385
MIP	2.250	.064	
Governing	–15.522	.000	
Agenda capacity	.652	.000	
IU			.054
MIP	–.146	.879	
Governing	3.944	.088	
Agenda capacity	1.104	.067	
CiU			.142
MIP	–2.598	.071	
Agenda capacity	3.789	.000	
PNV			.134
MIP	–.532	.523	
Agenda capacity	3.093	.000	

Note: Entries are the results of OLS regressions including a constant term (not shown). The dependent variable is the percentage of oral questions for each party that has regional content. The independent variables are the MIP (percentage of citizens considering nationalism as the most important problem, lagged one month), government-opposition status (scored 1 if the party is in government or supporting the government), and agenda capacity (which is the total number of questions posed by that party). The data cover the period of 1982 through 2011.

In both cases, the percentage of attention increases with increasing concerns about nationalism as the most important problem of Spain, and this is especially the case for the PP. Actually, most of the citizens that list nationalism as one of the three most important problems of Spain voted the PP in the past general elections, while only a small share of PSOE voters mentioned nationalism as the most important problem, and these cross-party differences are highly consistent over time (see *barométro del CIS*). Finally, agenda capacity has no impact on the level of politicization of political decentralization. After 2008, the number of questions asked by the PSOE declines dramatically but this seems unrelated to relative levels of attention to political decentralization. The same occurs in the case of the PP after 2012.

In short, statewide parties' attention to political decentralization varies across time, following a cyclical trend. Their main goal is to force the party in government to take a position about issues that are highly controversial, in which there is an important polarization of the electorate. On the other hand, regional parties always push regional issues in the agenda no matter who is governing and no matter what the state of national public opinion. These results also illustrate how the debate about political decentralization has increasingly become one of the preferred weapons of statewide political parties to attack each other in search for maximization of electoral rewards. We turn now to media coverage of the decentralization debate and whether it has contributed to the dramatic increases in polarization that we have already documented.

Media attention to political decentralization

Political decentralization captures an important share of the attention on the front pages of *El País* and *El Mundo*. From 1996 to 2000, most of these stories were about the capacity of the Basques and especially the Catalans to impose their demands on the minority government of José María Aznar (regions as veto players; see Figure 6.8). Though Rodríguez Zapatero also governed with a minority coalition, his dependence on regional parties was considerably lower, because his socialists were also governing in Catalonia in coalition with ERC and IU. After 2000, media attention was highly focused on various secessionist vindications in the Basque Country (*plan Ibarreche*) and the Catalan reforms of the Estatutos de Autonomía (Figure 6.9). Also from the late 2000s, attention to scandals of corruption at the regional level increased exponentially, with a series of scandals coming one after another (e.g., *caso Gürtel*, *caso Faisán*, *caso Palau*, and *ITV-Jordi Pujol* among others; see Figure 6.10). Finally,

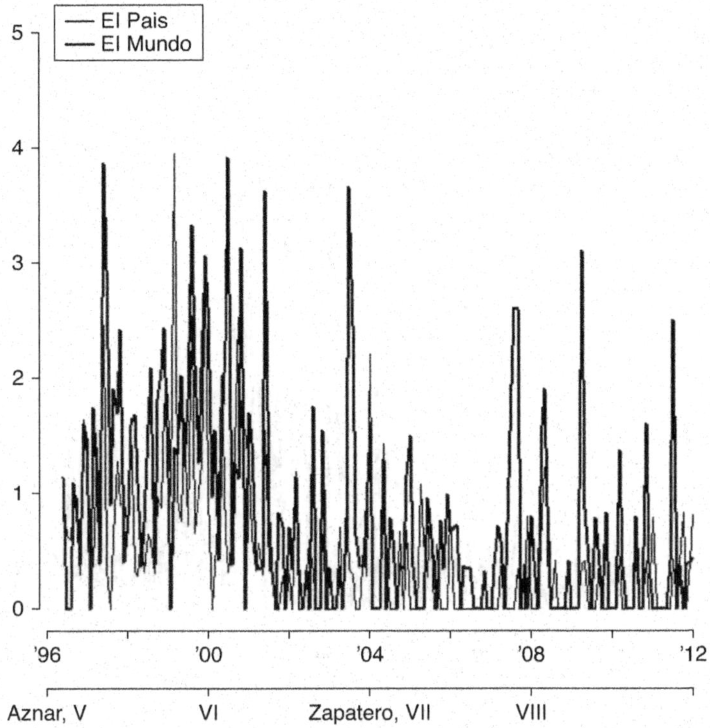

Figure 6.8 Attention to regions as veto players

Note: For each newspaper, the figure indicates the percentage of all stories mentioning the regions as veto players. Data are monthly.

Source: Spanish Policy Agendas Databases (www.ub.edu/spanishpolicyagendas).

attention to economic issues (regional fiscal policy), follow a U-shaped trend characterized by the fiscal reform in the 1990s and the debate about the regional fiscal balances, and regions' collective public deficit from the late 2000s (Figure 6.11).

Though the two newspapers we consider have different ideological tendencies, they have similar views toward the regional question; they support a unified nation. In general, they cover regional issues to a similar degree, as we can see by looking at the correlations in amounts of attention by month across the two papers; these correlations are high and significant for all these issues, especially in the case of corruption in the regions (r = .651), regions as veto players (.595), and the economy (.488); while in the case of secessionist vindications and political reform, correlation is lower but still significant (.235). One possible reason for these

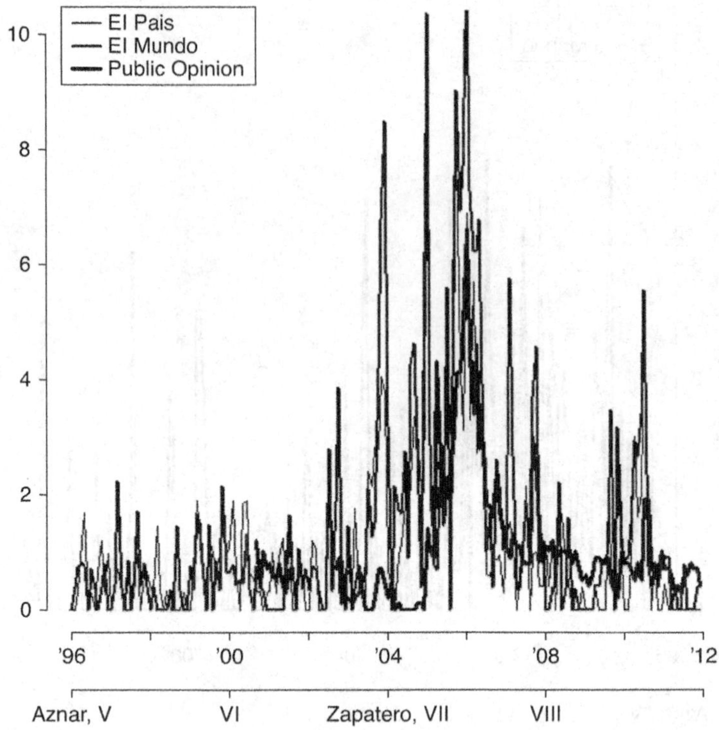

Figure 6.9 Attention to political decentralization in the media and public opinion

Note: For each newspaper, the figure indicates the percentage of all front-page stories on political decentralization. For public opinion, the figure indicates the percentage of all Spaniards indicating in national polls that political decentralization is the most important problem facing the country. Data are monthly.

Source: Newspaper: Spanish Policy Agendas Databases (www.ub.edu/spanishpolicyagendas). Public Opinion: Centro de Investigaciones Sociológicas (www.cis.es)

generally high correlations would be that the newspapers are simply following the actions of mainstream political leaders, for example, in the parliamentary debate. On the other hand, it could be that the political parties follow the media, focusing their parliamentary activities on regional issues only after the media have put these at the center of the national discussion and salience to voters.

To test these ideas, we develop two OLS models. In the first model the dependent variable is the percentage of media attention to regionalization by month and by political party, and the independent variables are (1) the percentage of oral questions about regionalization by month and

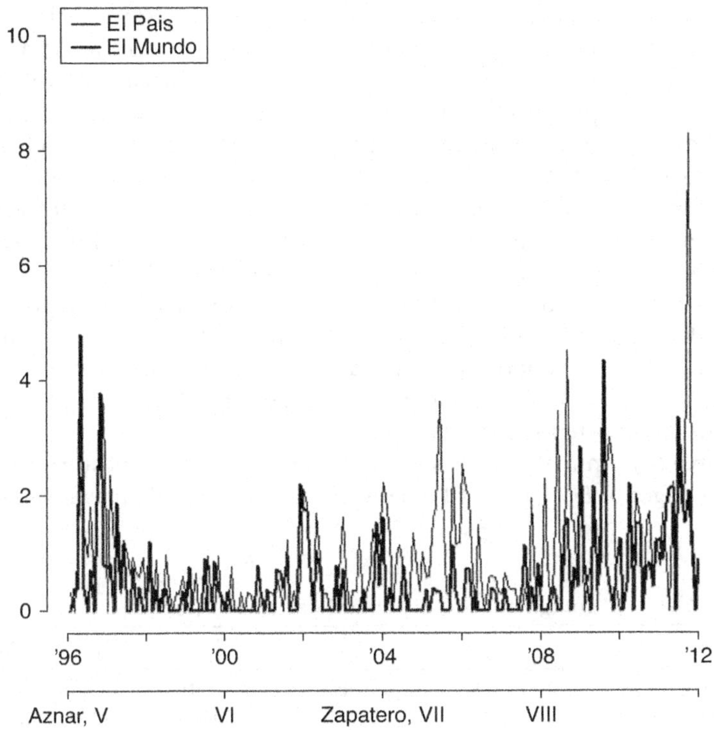

Figure 6.10. Attention to regional fiscal policy in the media

Note: For each newspaper, the figure indicates the percentage of all front-page stories on regional fiscal policy. Data are monthly.

Source: Spanish Policy Agendas Databases (www.ub.edu/spanishpolicyagendas).

by party, (2) the institutional position of that party in the government-opposition game (1 if the party is governing or giving support, and 0 otherwise), and (3) agenda capacity defined as the number of oral questions the party asked. In the second model, the dependent variable is the percentage of parliamentary questions by month and by party, while the independent variables are (1) the percentage of media attention by month, (2) agenda capacity, and (3) the institutional position of that party in the government-opposition game. In both cases the period covered is 1996 to 2011. Presenting two similar models allows us to see if questions lead the media or if the media drives the oral questions in Parliament.

As we argued in Chapter 5, we expect *El País* and *El Mundo* to follow parliamentary activities of the PP and PSOE, but not the smaller and regional parties. By the same token, we expect that political parties when asking questions tend to follow more than lead media stories. This is not that the media is setting the agenda about political decentralization. On the contrary, the media reports most of the time what political elites are saying or planning to do about this issue, giving special attention to the activities and statements of the Prime Minister and members of the government, and regional political elites. The media reports about these decisions, emphasizing those aspects that are more controversial and generate more debate. Political parties react to this information, asking questions that maximize their possibility of reelection, especially when they are in the opposition and they see the opportunity to erode popular support for the governing party. This leads us to expect that the main party in the opposition will react to the media with more intensity than others, and that regional parties will perform their tasks quite independently of what the media is doing.

Table 6.5 summarizes the results. First, the percentage of stories in the front pages of *El Mundo* and *El País* increase as MPs ask questions about

Table 6.5 Explaining attention to political decentralization

	B	Sig.	R Sq.
(a) Explaining oral questions about political decentralization			
Psoe Governing	3.344	.000	0.440
Agenda capacity	0.134	.000	
Stories in *El País*	0.800	.005	
Psoe Governing	3.435	.000	0.447
Agenda capacity	0.131	.000	
Stories in *El Mundo*	0.556	.007	
(b) Explaining stories about political decentralization in *El Mundo*			
Psoe Governing	0.945	.002	0.129
Oral questions	0.068	.007	
Agenda capacity	–0.004	.496	
(c) Explaining stories about political decentralization in *El País*			
Psoe Governing	0.755	.001	0.141
Oral questions	0.051	.005	
Agenda capacity	–0.007	.080	

Note: Entries are the result of OLS regressions including a constant term (not shown). We run three different equations: (a) the dependent variable is the percentage of oral questions of all political parties related to political decentralization. The independent variables are party in government, agenda capacity (which is the total number of questions posed by all parties, and the percentage of stories about political decentralization published in *El País*, or *El Mundo*. (b) The dependent variable is the percentage of stories about political decentralization in *El Mundo*; and (c) The dependent variable is the percentage of stories about political decentralization in *El País*. For cases (b) and (c) the independent variables are party in government, agenda capacity and the percentage of oral questions about political decentralization. The data cover the period from 1996 to 2011.

the issue, and especially when the PSOE is governing from 2004 to 2008; there is little difference between the two newspapers in these patterns. By the same token, oral questions (all parties) are affected by media attention and increase when the PSOE is governing, and especially when agenda capacity is larger (before 2008). Still, the coefficients are always higher and more significant when we analyze the impact of media in oral questions, rather than the other way around. Finally, when we look at the political parties separately, coefficients are only significant in the case of the PP, but not the other parties. That is, the number of questions about political decentralization asked by the PP increases as *El País* and *El Mundo* publish more stories about the regions, and this is particularly the case when the PP is in opposition. While these patterns are true for the PP, they do not hold for any other party in the Spanish political system. More than any other statewide party, the PP sets itself out as the primary opponent of regional vindications.

These results give further support to some of the arguments already discussed in Chapter 5. Political parties follow the media when they are in the opposition, with more intensity in the case of *El País* than *El Mundo*. Actually, *El Mundo* seems to follow the parliamentary agenda only for those issues related to the government (political decentralization) while, for the rest, the connection with parliamentary activities is much lower. Results for regional parties are not shown, because they are not significant for any of the cases. As we saw before, regional parties consistently ask questions about "their" issues, no matter what the role of the media.

Finally, Figures 6.11 and 6.12 show how the readers of *El País* and *El Mundo* are more and more exposed to negative stories about politics at the regional level. The number of stories about scandals of corruption increased exponentially from 2010 in both newspapers, with a high correlation between the two. Maybe even more relevant is the fact that the tone of the stories dealing with political decentralization has increased exponentially from the mid-2000s. Increasing negative tone about political decentralization – and increasing attention to corruption scandals at the regional level – contributes to the polarization and increasing confrontation in the Spanish democracy.

Conclusions

For the last decades, the capacity of the regions to govern within the geographical limits of their territories has increased significantly. In the early 1980s the issue of political decentralization was mainly about the distribution of issue jurisdiction and the construction and consolidation of basic institutions of this new model of the territorial distribution

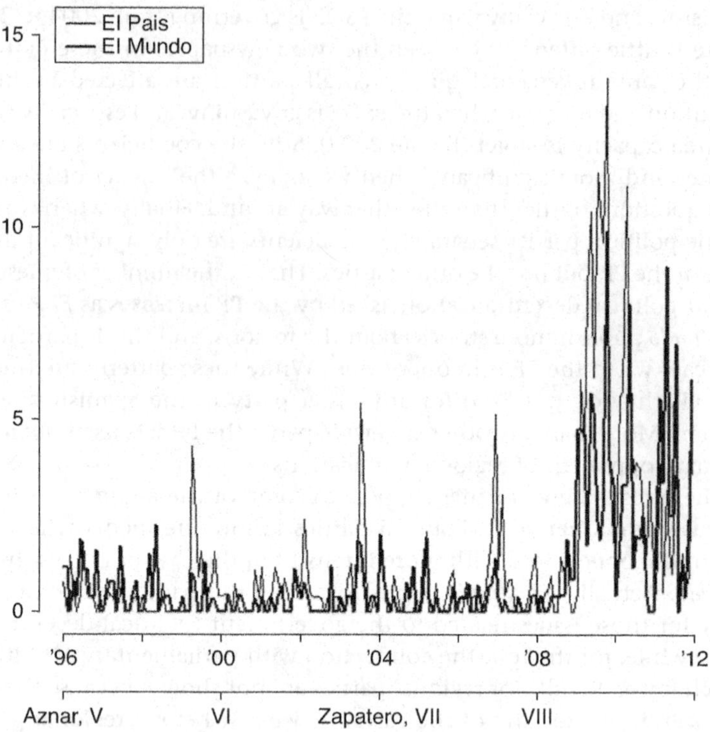

Figure 6.11 Political scandals at the regional level

Note: For each newspaper, the figure indicates the percentage of all front-page stories on political scandals at the regional level. Data are monthly.

Source: Spanish Policy Agendas Databases (www.ub.edu/spanishpolicyagendas).

of power. In the 1990s, policy change was designed first to guarantee the homogeneity of the process of political decentralization across the 17 regional governments, and later to increase regional fiscal autonomy. Finally in the 2000s, the debate was mainly focused on the reform of the Estatutos de Autonomía, and later on the vindications of political secession of some territories, especially the Basque Country and Catalonia.

The politics of decentralization have changed dramatically in the last decades toward increasing radicalization and confrontation in and out of the parliamentary arena. In this chapter, we explained how this process of radicalization of the political debate transcends the parliamentary arena, becoming a major object of citizens' mobilizations. In contrast to previous decades, the debate is not led by the minority governments of the PP or the PSOE vis-à-vis conservative regional parties, either CiU or

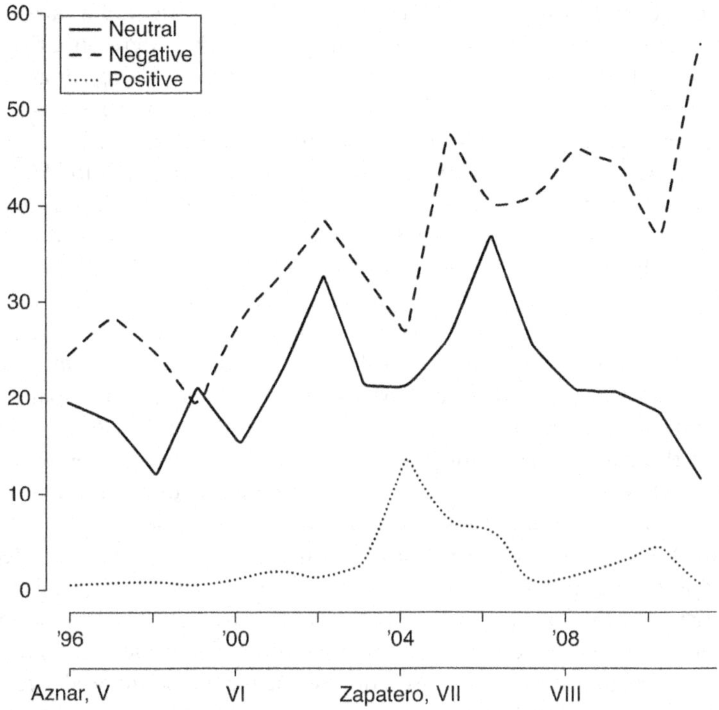

Figure 6.12 Tone of stories with regional content in *El País*

Note: Series indicate the percentage of stories that are positive, neutral, or negative in tone, as a percentage of all stories on the front page dealing with regions. Data are yearly. Positive refers to stories about increased powers for the regions or reporting good news about regional governments; negative, more powers for the central state or critical attention to regional governments.

Source: Spanish Policy Agendas Databases (www.ub.edu/spanishpolicyagendas).

the PNV, but rather by large social movements in alliance with political elites. By the same token, we have explained how the debate has moved from one centered on distribution of issue jurisdiction, to a debate centered on highly symbolic issues, such as the concept of nationhood. Further, the huge economic inequalities generated by globalization and other factors have been linked in the public debate with demands for regional autonomy.

Finally, we have explained how political elites have altered their parliamentary strategies, contributing to the decline of the politics of consensus and cooperation that dominated the debate of political decentralization in the 1980s, and especially the 1990s. Beginning around

2000, a steep decline has occurred on the adoption rate of parliamentary bills related to regional issues. Governing parties face increasing difficulties in finding allies in favor of organic laws related to regional issues and therefore cannot pass them. The large statewide parties serving in the parliamentary opposition increasingly use political decentralization as a means to erode the party in government. Both the PP and PSOE pay little attention to political decentralization when they are governing, but focus on it intensively when they are in the opposition. In doing so, they tend to follow public opinion, trying to maximize their electoral rewards. Our results also illustrate that regional parties follow a different pattern by pushing political decentralization into the parliamentary arena no matter which party is governing and regardless of public opinion. Finally, we demonstrate that the front pages of the two major newspapers in Spain have a significant and positive impact on the parliamentary activities of the two major political parties, but not regional parties. Also, we explain that newspapers are increasingly giving bad news about regional politics, and framing these stories using a negative tone.

There is no other issue in Spanish politics today that shows so clearly the increasing polarization and lack of consensus that characterizes Spanish politics in the last decade. This is completely the opposite pattern of the Europeanization of Spanish politics. European integration, and the delegation of issue jurisdiction to the EU, has generated a general agreement among political forces, and this is so despite the fact that the EU has imposed severe economic structural adjustments altering citizens' lives dramatically, and despite the fact that the EU has forced Spanish leaders to amend the Constitution in order to meet new goals relating to economic stabilization. Contrary to the politics of decentralization, and as we explain in the next chapter, the adoption rate of bills with EU content is larger than for the rest of bills, and political parties do not politicize EU matters as a means to erode the governing party. On the contrary, it is the party in government that highlights the EU as a means to maximize electoral rewards, with a few exceptions. The next chapter is designed to explain the politics of Europeanization. Spanish politics today is driven by alternating forces pushing power down to the regions and up to Brussels. The second transformation has had huge impacts on the Spanish public and yet it has been the object of almost no political controversy. Regionalization, on the other hand, has generated an entirely new politics of confrontation. We explore this puzzle by looking in more detail at the Spanish politics of Europeanization in the next chapter.

7 Europeanization

Contrary to the process of delegation of power downwards toward the regions, the transfer of political autonomy upwards toward EU institutions has not generated intense or highly politicized debate in Spain. Indeed, one of the main characteristics of the Spanish case is that party conflict over European integration and extreme euroskeptical parties are absent (Closa and Heywood, 2004; Arregui, 2007; Díez Medrano, 1995, 2003, 2007; Morata, 1998, 2004; Morata and Mateo, 2007; Piedrafita et al., 2006; Szczerbiak and Taggart, 2003; Kriesi et al., 2008; Hooghe and Marks, 2009; Rodríguez Aguilera, 2013, 2014). Although political parties can be critical of specific EU policies, they have neither questioned Spanish membership nor opposed the project of European integration. After Franco's death, identification with the EU became an important element in the political discourse of democratization, blurring ideological differences between political parties over EU issues, and making the emergence of euroskepticism unlikely (Closa, 1995; Quintanilla, 2000; Díez Medrano, 2003; Closa and Heywood, 2004).

Cross-party support for the EU has characterized the Spanish political party system, with the result that all EU treaties have been ratified with the broad-ranging support of all political parties and the Spanish population. Even the Constitutional treaty, where the leadership of the Far Left (IU and ICV) and the Catalan and Galician nationalists (ERC and BNG) asked their members to vote "no" because of its perceived minimization of national identities and neglect of social issues, had the support of more than 90 percent of the members of the Spanish Parliament, like it did the ratification of the Lisbon Treaty (Font and Rodríguez, 2006; Szmolka, 2007; Rodríguez Aguilera and Morata, 2008). The "no" supporters, in any case, were not against European integration but were asking for a "different" or "better" Europe (Torreblanca, 2005;

Backes and Moreau, 2012). Following the outbreak of the euro crisis, the Far Left (IU), that was already critical with the Maastricht treaty, has strongly criticized the subordination of the European project to the interests of the markets and its democratic deficit, but this does not justify characterizing this party as "anti-European" (Rodriguez Aguilera, 2014; Dunphy, 2004; Topaloff, 2012).

In line with the lack of party conflict, Spanish public opinion has always been more supportive and enthusiastic with regard to European integration than in other European countries (Szmolka, 1999; Sánchez, 1999; Díez Medrano, 1995, 2003, 2007). This positive attitude is attributed to the benefits that the country is believed to have gained from the EU at both the political and economic levels and mainly relies on the advantages that Europe offers, thought of in terms of better public policies and economic prosperity (Díez Medrano, 1995, 2007). This output-based form of EU legitimacy (Scharpf, 1999) is illustrated by the support that Spanish citizens give to different EU policies. The public has always been critical of the Common Agricultural Policy, which has more negatively affected Spanish interests, but has always been supportive of structural funds, as Spain has been one of the EU's principal aid beneficiaries until the entry of Eastern countries (Díez Medrano, 2007: 214). Citizens and parties do not question Spanish membership to the EU, but their support for European integration is strongly contingent on rational calculations (Díez Medrano, 2007).

Because the variables typically thought to account for the politicization of European integration are absent in the Spanish case, little attention has been paid to explain how the EU has affected domestic parliamentary debates. Even though important issues of public policy have been increasingly removed from the normal parliamentary scrutiny and debates and translated to a system dominated by the executive, we know little about how the EU has transformed government-opposition dynamics at domestic level. The case of Spain is excluded from most comparative analysis published on this topic (e.g., Holzhacker, 2002; Auel, 2007; Navarro and Brouard, 2012; de Ruiter, 2013; Finke and Dannwolf, 2013; Wonka and Rittberger, 2013; Auel and Raunio, 2014) and with some exception such as García -Lupato (2008),[1] existing literature mainly focuses on explaining the impact of Europe on particular policy areas (e.g., Torreblanca, 2001; Lombardo, 2003; Subirats et al., 2001; Chaqués-Bonafont, 2002; Palau, 2009); on formal rules and institutions like the Parliament or the executive (Arregui, 2007; Morata, 1987, 2004; Closa, 1995; Molina, 2001; Cienfuegos,

2001); on public opinion (e.g., Díez Medrano, 2003, 2007; Sánchez Cuenca, 2000); or on the electoral arena (e.g., Garcia-Lupato, 2008; Rodríguez-Aguilera, 2013).

This chapter is designed to fill this gap. To do so we analyze the EU content of all the oral questions introduced in the Spanish Parliament from 1986 to 2013. The goal is to explain the dynamics of parliamentary oversight on EU issues over time and whether there are differences between government and opposition groups. Results illustrate that even though important issues of public policy have been delegated upwards toward the EU, parliamentary groups do not use parliamentary question time to undertake routine oversight on EU decisions nor to compensate for their secondary role in the EU policymaking process. The lack of electoral incentives for politicizing European integration and the absence of euroskeptical parties explain why the EU is not among the top priorities of Spanish MPs. Nonetheless, Spanish MPs still frame questions in EU terms when it makes partisan sense to do so, and this is particularly linked to the government-opposition game. Their goal is not to politicize European integration but to use the EU frame in their competition with each other.

MPs with governmental status pay attention to the EU depending on the opportunities provided by the institutional context and the occurrence of EU events. They use parliamentary questions to explain the government's position in EU negotiations, to account for the results of EU meetings, or to explain the goals to be achieved during Spanish presidencies of the Council. The goal is to inform opposition groups, but also to give visibility to government's achievements and successes. Opposition groups use oral questions to promote debate on specific issues – especially those attractive to their voters – that would otherwise be ignored by the governing party. Given the output-based form of legitimacy that characterizes support for European integration, they also use critical junctures that occur when the proportion of citizens who believe that the country does not benefit from European integration increases to attach the government, portraying him as a poor negotiator, unable to defend national interests in Brussels.

The chapter is organized as follows. Section two describes attention to EU issues in the Spanish Parliament. Next, we explore what motivates government and opposition MPs to pay attention to EU affairs, considering the impact of institutional factors and EU events. The fourth section analyzes to what extent parliamentary groups use the EU frame following issue visibility strategies, and section five explains to what extent they react to variations in the benefits perceived from

European integration. The last part summarizes the main conclusions of the analysis.

Attention to EU affairs in the Spanish Parliament

Domestic parliaments have been mostly described in the literature as "losers" of the Europeanization process (e.g., Melsaeter and Sverdrup, 2004; Raunio, 2005: Saalfeld, 2000; Follesdal and Hix, 2006). Delegation of competences upwards toward the EU, has eroded its capacity to carry out some basic functions and weakened its role in the legislative process in favor of the executive. This is illustrated by the increasing Europeanization of the domestic legislative agenda (Brouard et al., 2012). In Spain, as Figure 7.1 shows, from 1986 to 2011, 36 percent of the total legislation passed in the Parliament has EU content.[2] The percentage of Europeanized legislation has progressively increased since Spanish entry into the EEC in 1986, especially since the realization of the Economic and Monetary Union, and focuses on specific policy areas, especially those related to the single market and the environment, where the percentage of EU-related legislation goes beyond 60 percent of the total (Palau and Chaqués-Bonafont, 2012).

The "de-parliamentarization" process is also illustrated by the increasing role of secondary legislation in the transposition of EU directives. Almost 90 percent of the legislation used to transpose EU directives is secondary legislation (mainly decrees – *Reales Decretos*) leaving a marginal role to the Spanish Parliament to carry out this task. The Spanish executive plays a leading role in the implementation of EU directives but also in the upstream process when EU legislation is discussed at the EU level. The government is designated as the sole representative in EU negotiations and monopolizes control over the agenda on European affairs. By contrast, the voice of domestic opposition groups is nonexistent or weak within EU institutions where most decisions are made in highly nontransparent Council meetings and the government can easily bypass domestic parliaments (e.g., Raunio, 1999; Follesdal and Hix, 2006; Holzhacker, 2002).

The leading role that the government has in the EU policymaking process and the high costs associated with the renegotiation of EU decisions at domestic level (Moravsick, 1994) explain why the adoption rate of bills with EU content is higher than those without EU content (92 percent vs. 89 percent), but not why bills with EU content have higher amendment rate (81 percent) than those without EU content (65 percent) (Palau and Chaqués-Bonafont, 2012). This is because the

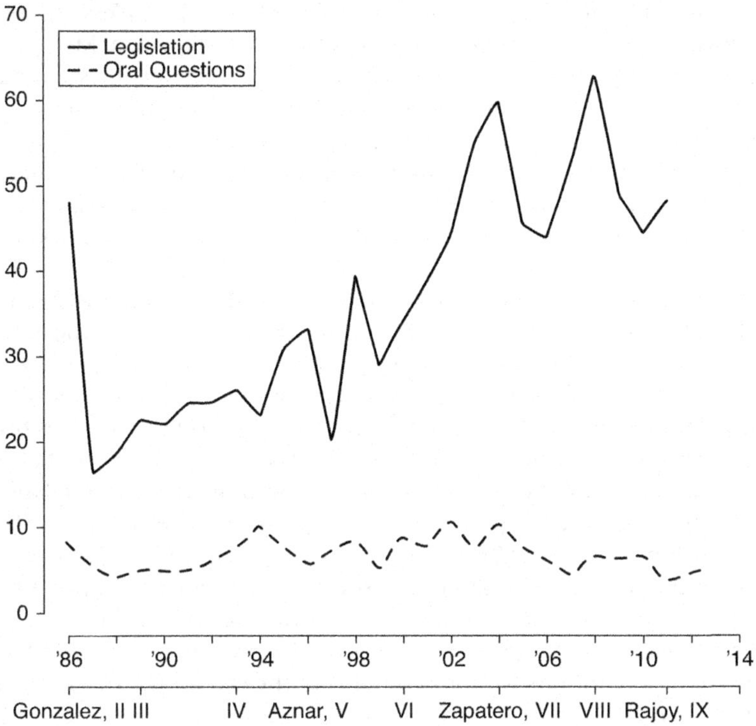

Figure 7.1 Percentage of legislation and oral questions that have European content

Note: The figure indicates the percentage of all legislation (ordinary laws, organic laws, decree-laws, and legislative decrees) or oral questions that have European content. Data are annual.

Source: Spanish Policy Agendas Databases (www.ub.edu/spanishpolicyagendas).

origin and motivations of amendments relate to factors other than Europeanization, from technical factors to ideological ones. In addition, the amendment rate is high and has increased for all types of bills, with and without EU content, which suggests that the amendment rate is probably related to other political factors in addition to Europeanization – like the increasing confrontation that has characterized Spanish politics in the last decade. Overall, the EU has contributed to reduce the already limited capacity of the Spanish Parliament to control the legislative agenda, confirming to a certain extent the arguments of those authors that support the idea that Europeanization involves a "de-parliamentarization" process of national political systems (e.g., Moravsick, 1994; Hix, 1999; Maurer and Wessels, 2001; Chryssochoou et al., 1998; Jachtenfuchs, 2001). Spanish MPs could use domestic parliamentary oversight to compensate for the secondary role of the Parliament and the gradual Europeanization of the legislative agenda, but the analysis of all the oral questions introduced in the Chamber from 1986 to 2013 illustrates that attention to EU issues in domestic oversight sessions is low and has not increased over time. As Figure 7.1 shows, only 7 percent of the total oral questions introduced in plenary meetings mention the EU.

The low saliency of EU affairs contrasts with that of other issues, like political decentralization, which is increasingly capturing the attention of Spanish MPs. The percentage of questions with regional content reached more than 17 percent of the total oral questions introduced in the Chamber from 2004 to 2008, while attention to EU issues has been relatively constant over time and never moved beyond 11 percent of the total. As Figure 7.2 shows, with the exception of international relations and agriculture, even in policy areas highly Europeanized the percentage of questions with an EU frame is very low. For example, 67 percent of environmental related legislation has EU content but only 8 percent of questions introduced in plenary meetings on this topic refer to the EU. The same applies to energy (55 percent of legislation Europeanized and 6 percent of questions with EU content) or the economy (53 percent vs. 6 percent).

The low saliency of EU affairs in the plenary is not compensated by increasing attention to the topic in more specialized committees. The percentage of questions with EU content introduced in parliamentary committees is also low and has not experienced any significant increase over time during the period 1986–2007. Not even the European Affairs Committee (EAC), the specialized committee designed to oversee government activity on EU affairs, has increased its scrutiny activity. In fact, the total number of questions asked in the EAC has declined over time, and especially following the creation of the EMU.[3] This committee has allowed opposition groups to gain access to EU-related information and to better control the executive's activity, especially after the reform introduced after the Lisbon Treaty,[4] but it plays only a modest role as it cannot limit the bargaining power of the government in front of EU institutions, as similar committees do in other countries such as Denmark or the UK[5] (Matia, 1999; Storini, 2005; Holzhacker, 2002: 435; Finke and Melzer, 2012).

The lack of electoral incentives resulting from EU-skeptic positions partially explains why the EU is not among the top priorities of Spanish

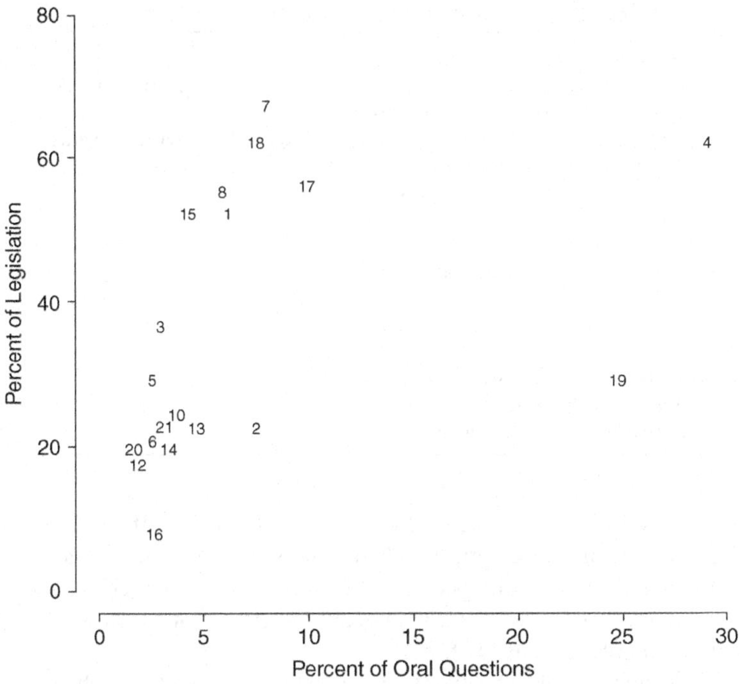

Figure 7.2 Comparison of EU content in oral questions and legislation, by topic area

Note: The placement of each number indicates the percentage of oral questions and legislation (ordinary laws, organic laws, decree-laws, and legislative decrees) on that topic that is Europeanized. Topics are listed by number as follows: 1. Macroeconomics; 2. Rights; 3. Health; 4. Agriculture; 5. Labor; 6. Education; 7. Environment; 8. Energy; 10.Transportation; 12. Crime and Justice; 13. Social Policy; 14. Housing; 15. Commerce and Banking; 16. Defense; 17. Science & Technology; 18 Foreign Trade; 19. International Affairs; 20. Public Administration, 21. Public Lands & Water Management. Topic 4 (Agriculture), for example, shows 30 percent Europeanization in the oral questions and 60 percent in legislation. Most issues are much more Europeanized with regard to legislation than in the oral questions posed in parliamentary question time.

Source: Spanish Policy Agendas Databases (www.ub.edu/spanishpolicyagendas)

MPs. However, the lack of party conflict on European integration cannot explain why certain issues have become increasingly "Europeanized" over time in plenary oversight sessions.

To analyze the content of EU-related questions we recoded all the oral questions introduced in plenary meetings according to the 13 macro categories described in Table 7.1. This slight revision to the standard CAP topic coding system allowed us to: (1) merge some CAP topic categories

Table 7.1 Main categories used for the analysis of Europeanization

1	Economic, fiscal, and monetary policy
2	EU funds (e.g., Cohesion and Structural Funds, European Social Fund, European Regional Development Fund)
3	Agriculture-related funds (e.g., European Agricultural Fund for Rural Development, European Agricultural Guarantee Fund)
4	Other agriculture-related issues
5	Fishing
6	Banking and commerce
7	Rights, justice, and internal affairs
8	Welfare
9	Labor, social security, and pensions
10	Environment and energy
11	Transport, housing, and communications
12	Defense and external policy
13	EU Treaties and institutions

with a very low number of oral questions with EU content, like social policy, health or education, which had large numbers of zeros; and (2) to consider relevant dimensions of EU politics not covered by the CAP topics and subtopics, which are especially relevant for the analysis of EU affairs in Spain, such as EU funds.

The proportion of EU-related oral questions across topics has significantly changed from 1986 to 2013. As Figure 7.3 shows, questions on agriculture, fishing, and EU funds represented over 60 percent of all questions with EU content during the 1980s, but today these issues account for only about 10 percent of the questions. This set of issues garnered a high level of visibility immediately after 1986, as the negotiations on these issues were among the most complicated conducted within the framework of Spanish accession (Ramon, 2007; Bassols, 1995). These negotiations induced an uptick in debate about domestic (and international) agricultural policy, like it did the successive reforms of the Common Agricultural Policy (e.g., the MacSharry or Fishler reforms passed in 1992 and 2003, or Agenda 2000). EU funds were also subject to high parliamentary attention during this period because of events like the reform of Structural Funds in 1989 or the creation of the Cohesion Fund by the Maastricht Treaty. As Figure 7.4 shows, however, attention to these issues has progressively declined in favor of new issues that have entered the EU parliamentary agenda during the last decade.

Figure 7.4 shows that economic, fiscal, and monetary policies have increasingly been the focus of the Spanish debate on EU issues. Increases and decreases in attention to these topics clearly illustrate the impact

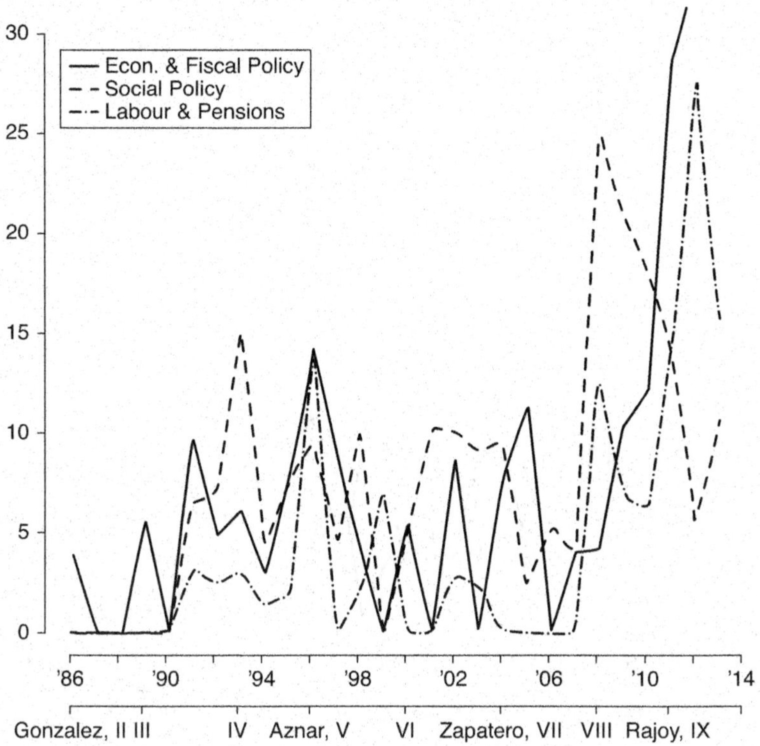

Figure 7.3 Topics with rising EU-related attention

Note: The figure shows the percentage of oral questions with EU content for three policy domains where this percentage is rising over time as a percentage of all of those that deal with European issues. Data are annual.

Source: Spanish Policy Agendas Databases (www.ub.edu/spanishpolicyagendas).

of the Maastricht treaty, the fulfillment of the convergence criterion and the creation of the EMU, or the adoption and modification of the Growth and Stability Pact. The EU frame of economic affairs increased significantly following the outbreak of the economic and financial crisis in 2008 – almost 30 percent of the total EU agenda is related to macroeconomics. Trailing behind the increased attention to economic issues, social issues and pensions and labor issues have also progressively entered the agenda after 2008 as a result of the policies implemented to overcome the crisis following recommendations of EU institutions, including the pension reforms implemented by Zapatero in 2011. Although these represent a small percentage of the total EU agenda

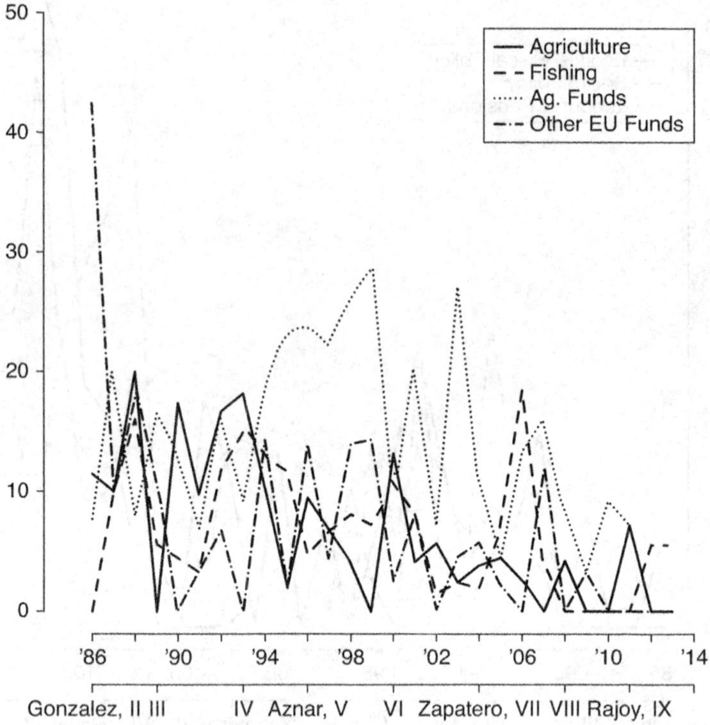

Figure 7.4 Topics with declining EU-related attention

Note: The figure shows the percentage of oral questions with EU content for four policy domains where this percentage is declining over time as a percentage of all of those that deal with European issues. Data are annual.

Source: Spanish Policy Agendas Databases (www.ub.edu/spanishpolicyagendas).

(8 percent and 3 percent, respectively) they are increasingly capturing the attention of Spanish MPs.

Finally, there is a set of EU issues whose attention has not been subjected to any substantive variation over time. This is the case of rights and justice and defense, which are rarely framed in EU terms (7 percent and 6 percent, respectively), given the intergovermental approach that still characterizes these policy areas. Attention to EU treaties and negotiations is contingent upon the occurrence of EU events, like the signature of EU treaties and Spanish presidencies of the Council, and represent an important part of the total EU agenda (12 percent). Environment and energy and industrial policy (6 percent and 3 percent of attention, respectively) fall also within this category of issues, like transport and communications, which represent 12 percent of the total EU issues. This high percentage is explained by the intense debates generated by the EU investment decisions on this policy – like the Mediterranean Corridor, or because Spain has a railway track gauge different from that of the main rail network within the Union – which have important implications for communications and trade. While these topics sometimes involve the EU to a significant extent, we see no important trends over time in the degree to which it is the case; these are constant if minor elements of the EU as reflected in Spanish parliamentary questions.

In short, overall parliamentary scrutiny over EU issues is low, but important issue dynamics affect Spanish consideration of the EU. In the following sections, we turn to analyze to what extent these dynamics are related to the government-opposition game.

Government-opposition dynamics

Parliamentary oversight sessions on EU issues are mainly dominated by the government. Even though the government holds superior information on any specific EU decision and law proposal, opposition parliamentary groups do not introduce more questions with EU content than the government to reduce this information asymmetry nor to compensate for its secondary role in the EU decision-making process. As Figure 7.5 illustrates, for most of the period from 1986 to 2013, government MPs have introduced a higher percentage of oral questions with EU content in plenary meetings than those in opposition. On average, governing parties devote 9 percent of their parliamentary oversight agenda to EU affairs; those in opposition, 6 percent.

The lack of party conflict over European integration and the difficulties for effectively influencing the executive's position in supranational negotiations mean that opposition MPs decide to allocate attention to other (domestic) priority issues. This results in the government paying more attention to EU issues than opposition groups do with important differences between the two main parties that have been in office in Spain: the PP and the PSOE. The PP introduces more questions with EU content when it is governing (11 percent) than the PSOE (7 percent). This difference can be partly explained by institutional factors and more specifically by variations in agenda capacity described in Chapter 4.

Even though Europe was one of the main priorities for the socialist government of Felipe González, who perceived national and European

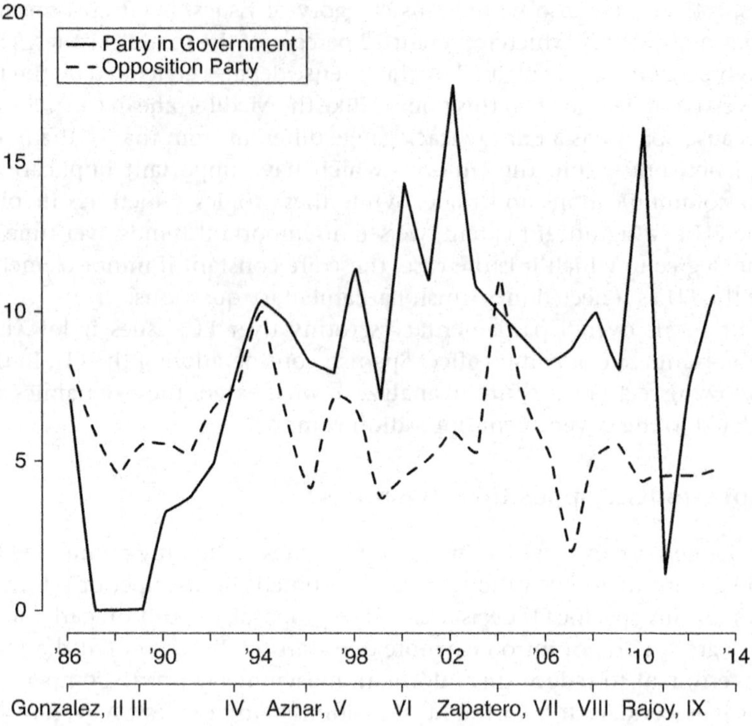

Figure 7.5 Percentage of oral questions with EU content registered by the government and opposition parliamentary groups

Note: The figure shows the percentage of oral questions with EU-content posed by members of Parliament from parties in the government or in the opposition.

Source: Spanish Policy Agendas Databases (www.ub.edu/spanishpolicyagendas).

interests as strongly intertwined (Barbé, 1999; García-Lupato, 2008), the socialists only occasionally referred to the EU in oversight sessions before Maastricht. González devoted an important part of attention to Europe in his parliamentary speeches during the 1980s, but this saliency was not translated into oversight sessions. Figure 7.6 shows the total number of questions posed by government MPs in oral question time and the percent of those questions with EU content. The low capacity of the Chamber to process questions and the advent of a difficult political situation in the early 1990s, with rising unemployment rates and the government involved in different corruption scandals, prevented the entry of EU issues into the parliamentary oversight

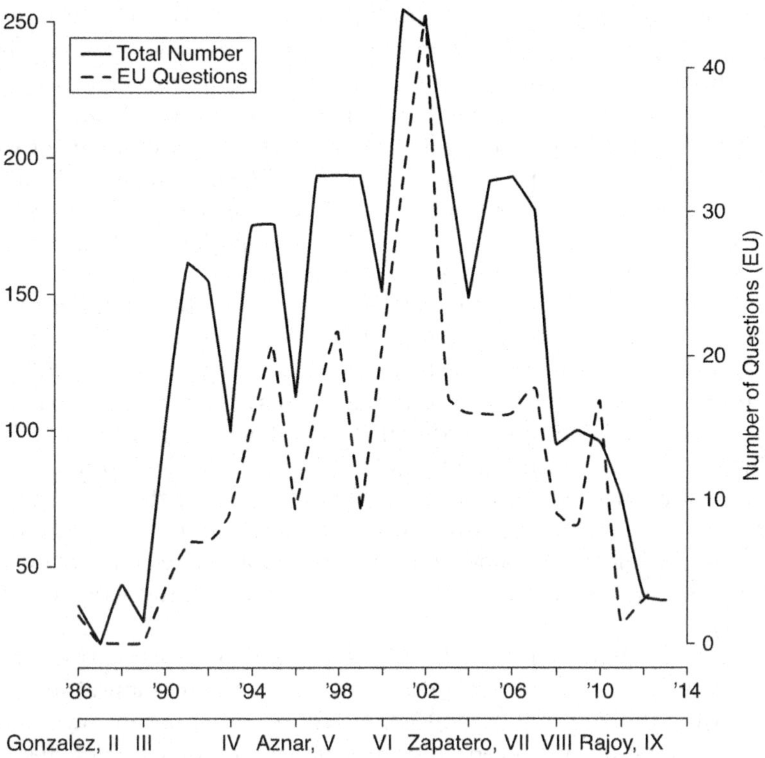

Figure 7.6 Agenda capacity and government attention to EU affairs

Note: The figure shows the total number of questions asked in parliamentary question times (left scale) and the number of questions with EU-related content (right scale). Data are annual.

Source: Spanish Policy Agendas Databases (www.ub.edu/spanishpolicyagendas).

agenda during these legislatures. This occurred even though Europe was a top priority for the socialist government and important events related to the EU occurred during this period (e.g., the Spanish presidency of the European Council in 1989, the reform of EU funds or the discussion of the single market).

Government attention to EU issues in parliamentary debates increased during the Aznar legislatures even though, beyond the creation of the EMU, Europe played a less important role for the conservatives (Barbé, 1999; García-Lupato, 2008). Variations in agenda capacity partly account for the higher attention that the PP pays to EU issues compared to the PSOE. In all parliamentary groups there is a positive correlation between agenda capacity and attention to EU affairs: the more questions asked in the Chamber, the higher the percentage of EU-related questions.[6] This suggests that EU-related questions are seen as somewhat superfluous to the core political issues of concern to Spanish MPs; if the available space is highly limited, very few EU questions are posed, and these numbers increase substantially only when the total number of questions posed moves to higher levels. But the relationship between agenda capacity and attention to EU affairs is especially high in the case of the PP. This is explained because the number of questions registered and asked became more predictable after 1996 when parliamentary rules first defined a fixed number of questions per session and parliamentary group. This increased the incentives of government MPs for using its allotment quota of questions, as otherwise rival parties had the opportunity of asking more questions attacking the government (see Chapter 4). By framing questions in EU terms, the government gave visibility to successful policies, like the fulfillment of the convergence criteria, the entry into the third stage of the EMU or the introduction of the euro, while at the same time used its quota of questions, reducing the opportunities of opposition groups to investigate the actions of the executive.

Governmental attention to EU affairs can be partly explained by formal rules, but it is also contingent on the occurrence of relevant EU events. Following the creation of the EMU, attention to European integration decreased significantly even though parliamentary rules did not change. In 2002, the PP devoted an important part of its oversight agenda to EU issues mainly because the introduction of the euro concurred with the Spanish presidency of the Council. Similarly, the high attention to EU affairs in 1995 is not explained by institutional factors but by the Spanish presidency that occurred during that year. The same can be argued for the rising levels of attention to European integration during the negotiations and signature of the Amsterdam, Nice or Lisbon treaties in 1997, 1999, and 2007, respectively.

To test to what extent MPs react to EU events, following Guinaudeau and Palau (2013), we have created a categorical variable signifying the number of relevant EU events that occur in a given month. We considered the following EU events: the signature of EU treaties, including enlargement treaties and referendums; European Parliament elections; European Council summits; EU Spanish presidencies; and key events related to the EMU, like the introduction of the euro or the publication of convergence reports (see Appendix D, this volume). Then, we ran two different regression models with the percentage of oral questions with EU content as the dependent variable, controlling for differences between government and opposition, and including a lag variable to test the autoregressive character of the series.

Regression results illustrate that parliamentary attention to EU affairs increases when there are EU events due to the increased attention from government MPs but not of opposition MPs (Table 7.2). This gives support to the idea that the government uses parliamentary questions to explain the Spanish position in EU negotiations, to account for the results of EU meetings and summits, and/or to explain the goals achieved during Spanish presidencies. In sum, government MPs, but not opposition ones, pose questions designed to draw attention to the government's policy achievements in Brussels. Opposition parties do not contest, but they do not draw attention to this as they have no political or partisan incentive to draw negative attention to the EU, as all parties strongly support it. The only exception was the Maastricht treaty and the Constitutional treaty that gained attention among left and regional groups for their lack of social content and neglect of regional interests.

During the Zapatero legislatures, government MPs paid less attention to EU affairs compared to Aznar even though important events occurred during this term for different reasons. First, in contrast to what happened with the EMU, the treaties signed during Zapatero's presidency could not be framed as a governmental success. The European Constitution project found the opposition of IU, ERC, and the BNG, and the project collapsed altogether when most countries failed to ratify it. Second, the signature of the Lisbon treaty and the Spanish presidency of the Council

Table 7.2 Impact of EU events on the percentage of oral questions with EU content introduced by government and opposition MPs

Oral questions with EU content	B	Sig.	R^2
Government			.025
Lagged questions	.002	.974	
EU events	2.216	.024	
Opposition			.001
Lagged questions	–.021	.766	
EU events	.148	.803	

Note: OLS regression results are shown. The dependent variable is the percentage of oral questions with EU content introduced by government and opposition groups. The independent variables are the percentage of questions with EU content lagged one month and the occurrence of EU events (see the text for details). Data are monthly and cover the period from 1986 to 2013. The models also include a constant term (not shown).

in 2010 raised attention to EU issues, but the outbreak of the crisis that led to an increasing unpopularity of EU institutions decreased the incentive of government MPs to frame questions in EU terms. The delegation of decision-making powers to the EU and the creation of the EMU meant that the government had to implement unpopular policies to curb spending and to control the state deficit in line with EU recommendations. The reform of the Spanish Constitution to cap budget deficit or the reform of the pension system in 2011 are some examples. Finally, a new reform of the parliamentary rules passed in 2008 (see Chapter 4) again decreased the incentives of government MPs to ask questions in the Chamber, reinforcing the already low incentives they had to pay attention to EU affairs.

The main conclusions to be drawn from the above are that attention to EU affairs is mainly dominated by parliamentary groups associated with the government, and these groups pay attention to the issue depending on the opportunities provided by the institutional framework and the occurrence of EU events. The rest of the chapter is devoted to explaining the motivations of opposition parliamentary groups to pay attention to EU affairs. We start by explaining that, just as the government uses the EU to spotlight successful policies, the opposition parliamentary groups use the EU to give visibility to bad news and issues that would otherwise be ignored by the governing party.

Issue visibility strategies

Regional political parties, such as CiU, PNV or ERC, introduce more oral questions with EU content than other opposition groups, especially compared to the Far Left (IU). ERC and CiU each introduced 8 percent of questions with EU content, and the PNV introduced 7 percent, while the IU was lower at only 5 percent. Contrary to what we observed in Chapter 6 regarding political decentralization, the high visibility of EU issues among regional parties is not related to the type of government. The minority governments of González and Aznar oversaw the implementation of the Maastricht treaty and the fulfillment of the convergence criteria; and most stages of the implementation of the EMU took place during this period, which made these governments especially relevant to the development of EU policies. Although CiU and the PNV exerted greater influence during these minority governments, they did not introduce more oral questions with EU content under minority than when under majority governments. CiU introduced 8.3 percent questions with EU content when it does not give support to the central government and 8.4 percent when it is pivotal;

and the PNV introduced 7 percent and 5 percent, respectively, differences that are low and not statistically significant.

This result is unsurprising, because both regional and governmental parties shared common goals regarding EU policies during this period. As Garcia -Lupato (2008) illustrates, during the socialist minority government, the CiU supported the socialist government, alongside its pursuit of political stability, because they supported the creation of the EMU, which the socialist government initially oversaw. During Aznar's first term, the regional parties provided strong support to the government in order to facilitate the implementation of the Maastricht criteria, and they minimized discussion about the procedures and form related to the implementation of the treaty.

According to the issue-competition perspective (Green-Pedersen and Mortensen, 2010; Green-Pedersen, 2012), this high percentage of attention devoted to EU issues by regional groups can be explained because they use the EU frame to promote debate on specific issues, lending visibility to regional interest and demands. This is illustrated by the large percentage of oral questions with EU content that mention also regional interests: 67 percent in the case of the PNV, 29 percent for CiU, and 44 percent for ERC. In the case of statewide parties this percentage decreases to 4 percent (PP) and 3 percent (PSOE). By framing regional issues in EU terms these parties give visibility to issues that would otherwise be ignored by the incumbent and to interests and demands that from their point of view cannot, or are not, defended adequately by the governing party in front of EU institutions.

A large proportion of the EU-focused questions asked by Basque and Catalan MPs is designed to advocate for increased representation and participation of regional authorities at the EU level or to draw attention toward language issues – urging the government, for example, to take actions to facilitate the recognition of the Catalan or Basque language in EU institutions. Regional groups are also especially concerned with what their territories get from European integration, especially in terms of EU funds or benefits obtained from Common Agricultural Policy reforms. An important part of their attention focuses also on transport and communication issues, especially to EU support for autonomic TV stations and EU funds designed to support the development of infrastructures at regional level.

Overall this illustrates that regional parties, and especially the PNV, pay particular attention to the interests of their constituencies also when asking EU-related questions. The question we ask now is to what extent regional and also other opposition groups react to critical junctures occurring when the benefits perceived from European integration decrease. Regional groups have incentives to pay attention to the EU in these situations to show their voters that they care about the interests of their territories, but other opposition groups may also use this negative environment to attack the government as part of the government-opposition game.

Reaction to critical junctures

To test to what extent Spanish MPs react to variation in public opinion perceptions of European integration we have used Eurobarometer data series and oral questions to run two regression equations. We test whether variations in the "benefits perceived from EU integration" explain changes in the percentage of oral questions with EU content introduced by parliamentary groups and whether there are differences between government and opposition MPs. Because the Eurobarometer series is only collected semi-annually we could not develop a single regression model, testing the variables considered in the previous section and the impact of public opinion at the same time. Here, we have considered to what extent variations in public opinion affect the percentage of oral questions with EU content introduced between Eurobarometer polls for the period from 1986 through 2011.[7]

The results of the regressions (Table 7.3) illustrate that variations in public opinion do indeed drive parliamentary behavior but with important (statistically significant) differences between government and opposition groups. When the percentage of citizens considering that Spain does not benefit from European integration increases, opposition

Table 7.3 Variations in the percentage of oral questions with EU content based on variations in the benefits perceived from European integration

Oral questions with EU content	**B**	**Sig.**	**R^2**
Government			**.233**
Public opinion	–.180	.000	
Opposition			**.112**
Public opinion	.072	.019	

Note: OLS regression results shown. The dependent variable is the percentage of oral questions introduced with EU content by government and opposition groups. Public opinion refers to the percentage of citizens considering that Spain does not benefit from European integration. Data cover the period between 1986 through 2011 and include observations whenever there is a Eurobarometer poll, which provides the public opinion data. Generally, this is every six months.

parliamentary groups intensify their scrutiny over EU issues. But government MPs do not react similarly to variations in public opinion. In fact, they ask fewer questions with EU content in the context of lower public opinion toward the EU. These statistical results quickly summarize the historical and political relationship between public opinion on the EU and the use parties have had for the EU frame.

As Figure 7.7 illustrates, there have been important variations in the benefits perceived from European integration over time. In mid-1980s,

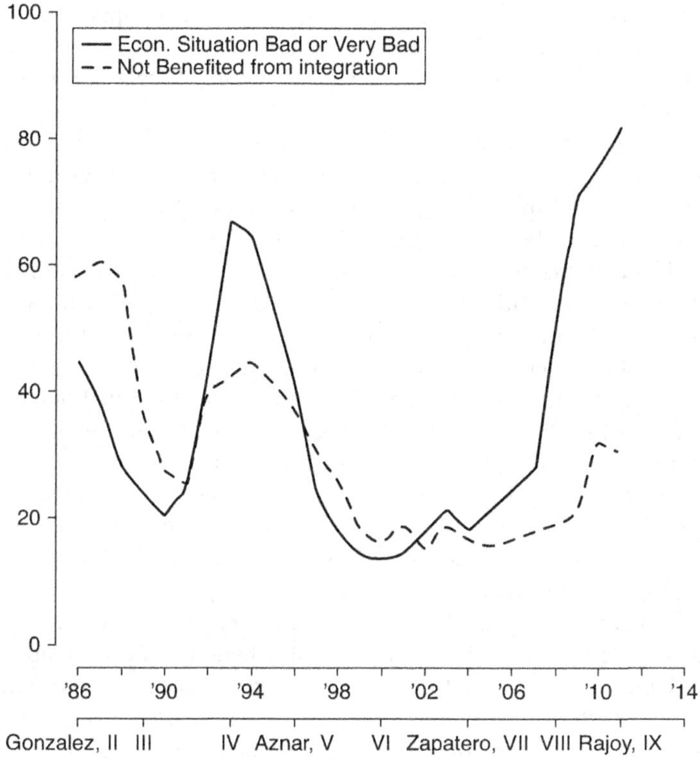

Figure 7.7 Public opinion perception of the economic situation and benefits perceived from European integration

Note: The figure shows the percentage of public opinion responses saying the economic situation is bad or very bad and the percentage saying that the country has not benefited from European integration. During bad economic times, the perceived benefits of the EU appear to decline. Data are annual.

Source: Public opinion about the economy: Centro de Investigaciones Sociológicas (www.cis.es). Public opinion about the benefits perceived from European integration: Eurobarometer (www.ec.europa.eu/public_opinion/index_En.htm).

only 15–20 percent of citizens considered that, on balance, the country had benefited from being a member of the European Community (EC). This is explained because, as Jones (2000) points out, right after Francoism and during negotiations with EU institutions, there was a strong political consensus about the desirability of accession. At this point, the central government had the ability to convince vulnerable sectors that they would ultimately benefit from accession, which inhibited the mobilization of nongovernmental actors during the process. The fragmentary nature of lobbying in agricultural and industrial sectors strengthened the central government capacity to avoid opposition to EC entry, that was perceived as a way to strengthen democracy and improve the Spanish position in the international system (Díez Medrano, 2007; Jones, 2000).

Initial support for European integration was a diffuse and emotional reaction to the values that Europe represented rather than utilitarian in nature; membership was supported more for political reasons than for specific economic benefits (Barbé, 1999: 20). Opposition groups reacted to the low benefits perceived from European integration by asking questions related to agriculture, fisheries, and EU funds – policy areas where the impact of Europe was perceived as more negative for Spanish interests. These questions were mainly introduced by the PP (at that time in opposition) and by regional parties that were especially worried about how EU funds would be distributed among the regions, and for the impact that the Common Agricultural Policy could have on their territories, especially regarding those products on which their regions had competitive advantages. As Table 7.4 illustrates, agriculture and other EU-related funds are the main priorities of regional parties (CiU and PNV).

The capacity to obtain positive returns for Spain in the 1988 negotiations regarding the EU budget and structural funds was used by the socialist government to illustrate the visible benefits of EU membership. From 1987 to 1993, structural funds increased from approximately 7 billion to 14 billion ECU (Jones, 2000: 89). As a result, and also coinciding with an overall improvement of the economic situation at the domestic level, the benefits perceived from European integration increased in the late 1980s until Maastricht, when they suffered an important decline. Pessimism accompanying the ratification of the Maastricht treaty, the economic crisis of the early 1990s, high unemployment rates, increased commercial deficits, and high levels of political instability explain this decline that occurred in the early 1990s (Díez Medrano, 2007).

Opposition groups reacted to these depressed evaluations of benefits by shifting attention from agriculture, fisheries, and EU funds to

Table 7.4 Percentage of oral questions with EU content by parliamentary group, governing status, and topic

	PSOE		PP				
	In opposition	Governing	In opposition	Governing	IU	CiU	PNV
Total	**5**	**7**	**5**	**11**	**5**	**8**	**7**
By policy topic:							
Economic, fiscal, and monetary policy	14	2	10	5	18	3	4
EU funds	10	2	6	4	2	7	24
Banking and commerce	1	3	3	5	2	8	0
Rights, justice, and internal affairs	3	13	6	6	7	8	7
Other agriculture-related issues	3	7	10	6	7	12	0
Agriculture-related funds	16	14	16	17	9	18	4
Fishing	2	10	7	7	2	0	11
Welfare	7	15	4	9	4	10	2
Labor, social security, and pensions	5	2	4	2	2	4	7
Environment and energy	11	7	7	4	9	4	2
Transport, housing, and communications	7	8	9	18	13	12	22
Defense and external policy	8	7	4	8	2	3	4
EU Treaties	11	11	12	10	22	11	13
Total EU-related questions	100	100	100	100	100	100	100

Note: Covers the period of 1986 through 2013.

Source: Spanish Policy Agendas Databases (www.ub.edu/spanishpolicyagendas).

economic and monetary issues. As Figure 7.8 illustrates, the main opposition groups increased the percentage of questions introduced on EU economic and monetary affairs from 1992 to 1995, when the socialists were governing. These questions were not asked by regional parties but mainly by the PP and IU. The conservatives were especially concerned with how far the Spanish economy was from fulfilling the convergence criteria defined in the Maastricht treaty and by the appropriateness of the policies adopted by the socialists in order to enter the EMU. IU intensified its scrutiny on EU issues during this period mainly because of concern about how the government would make compatible the entry into the EMU with the maintenance of social programs. This explains

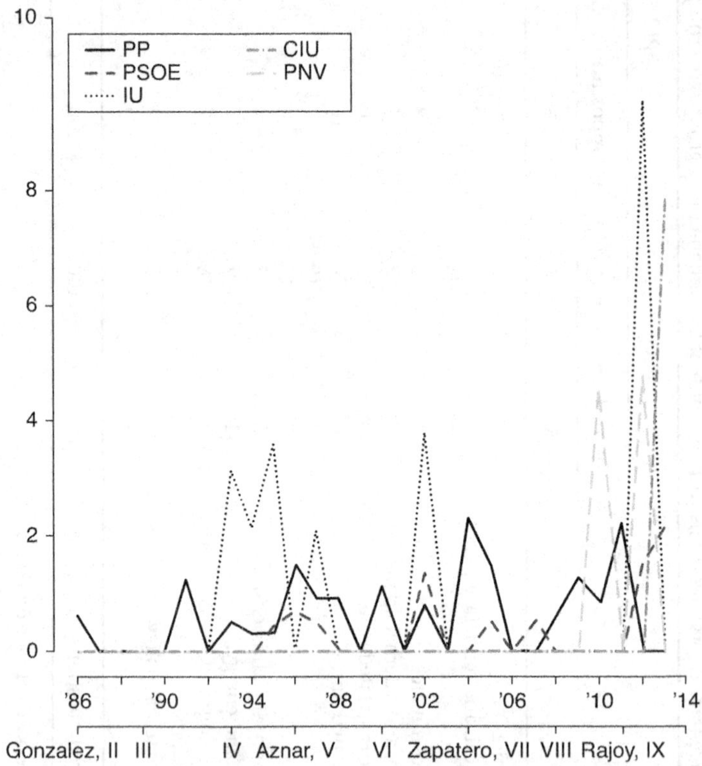

Figure 7.8 Attention to EU economic, fiscal, and monetary policy by party group

Note: The figure shows the percentage of all EU-related oral questions posed in Parliamentary question time that relate to EU economic, fiscal, and monetary, by party group.

Source: Spanish Policy Agendas Databases (www.ub.edu/spanishpolicyagendas).

why, contrary to the conservatives, this party framed an important part of welfare-related questions in EU terms in the early 1990s. However, as Table 7.4 illustrates, IU is not the group that pays more attention to welfare issues, as the PP and the PSOE give them more visibility – although only when they are governing. CiU devotes also an important part of its EU questions to welfare, although in this case, attention focuses on education, culture, and language, which are also coded in this category.

As Table 7.4 shows, IU, the PP, and the PSOE, when in opposition, are the groups that pay more attention to EU economic and monetary issues, although they do so with quite different objectives. Once the PP reaches power in 1996, it mainly uses the EU frame to illustrate the success of the government's economic policy and to justify policy decisions like a new labor-market reform introduced in 1997 or the liberalization of key sectors of the economy. The IU, by contrast, stresses the desirability of making the convergence criteria more flexible, especially regarding public expenditures, and to make them compatible with policies to promote job creation, which explains the high percentage of attention devoted by this group to EU labor issues in the late 1990s. Once in opposition, the PSOE also uses the EU to attack the government, following the same strategy previously used by the PP. Regional parties, on the contrary, do not frame economic affairs in EU terms. Neither CiU nor the PNV introduced any questions related to the EU economic and monetary policy during the 1990s when they were giving support to the minority government of the PP.

During the late 1990s, when the economic situation recovered and when fears that Spain would be unable to fulfill the convergence criteria vanished, the benefits perceived from European integration progressively increased. Following the completion of the EMU, there was a little step backwards because of the rise in inflation caused by the introduction of the euro; but since 2002, the perceived benefits stabilized, coinciding also with a period of good economic performance. As Figure 7.7 illustrates, there is a high correlation between domestic economic performance and the benefits perceived from European integration. This supports the idea that attitudes at the domestic level spill over onto EU institutions (Anderson, 1998). As Spanish citizens' support for the EU is based on a rational economic calculation and they are largely ignorant about EU affairs, they do not develop an independent judgment about the EU, but rather their attitudes toward European integration are strongly contingent upon the domestic economic situation (Díez Medrano, 2007).

Overall, from the introduction of the euro to the outbreak of the current economic crisis, a very low percentage of citizens considered that the country was not benefiting from European integration, even though during this period important decisions were made with implications for Spanish interests, like the 2003 Common Agricultural Policy reform or the negotiation rounds for the 2007–2013 EU budget that envisaged Spain as a net contributor to the EU budget for the first time. These events did not have a strong impact on public opinion because they occurred in a favorable economic context while the government was very popular. In the parliamentary arena – with the exception of CiU, which asked some questions about the impact of the Common Agricultural Policy reform in Catalonia, and the PP regarding the budget negotiations – they did not arouse particular interest on the part of most MPs, who were more concerned about the ratification process of the EU Constitution, especially IU and ERC. The opposition of the Far Left to the Constitutional treaty, the Lisbon treaty, and also the concerns it raised over the Maastricht treaty in the early 1990s, explain why IU devotes more attention to EU treaties than other parliamentary groups do (see Table 7.4).

In 2008, following the outbreak of the economic crisis, the percentage of citizens increased who gave consideration to the idea that the country does not benefit from European integration. The euro crisis has generally exposed a weakness in the output-directed legitimacy of the EU (Jones, 2009). In the case of Spain it has significantly affected one of the pillars that has traditionally sustained support for European integration: economic growth and the reduction of the difference in living standards between Spain and the rest of Europe (Díez Medrano, 2007). Unemployment rates have almost tripled since the beginning of the crisis, from 9 percent of the workforce in 2007 to 22 percent in 2011 and 27 percent in 2013. The Spanish economy entered recession in 2009, with a 4 percent negative growth rate, while rising public and private debt levels led to a serious sovereign debt crisis. Family incomes have fallen to the levels from ten years ago, putting three million people in extreme poverty conditions. This poor economic performance and the social consequences of the policy measures adopted to combat the crisis, most of them recommended by EU institutions, undermined seriously the popularity of the incumbent governments and also that of EU institutions. In 2013, a large majority of Spaniards (86 percent) thought that if EU institutions continued to implement the policies that had been developed to manage the crisis, the EU would no longer be "useful" and 94 percent considered it necessary to reform them (according to a CIS poll conducted in 2013).

This context was used by opposition groups to attack the government, and this was particularly the strategy followed by the PP, given its high probabilities to get into office in the next elections. The conservatives asked questions with EU content not to criticize EU policies but to embarrass the Zapatero government, urging it to accomplish policy reforms in line with the recommendations made by EU institutions. For example, before a new pensions law was passed in 2011, the conservatives asked the socialists why they were not taking into account the Commission's recommendation regarding this policy to face the challenges of an aging population. Similarly, when the housing bubble broke and the economy collapsed, they asked the government how it was going to deal with the bad economic projections made by EU institutions for Spain. This strategy explains why the PP, from 2008 to 2011, before taking office, increased the percentage of questions related to the EU economic and monetary policy (Figure 7.8) and also to EU labor and pensions-related issues.

The victory of the PP in the 2011 general elections by an absolute majority initiated a new period characterized by higher levels of confrontation and during which all opposition groups increased their attention to EU economic and monetary policies. IU, which had not asked any question related to these policies during the second legislature of Zapatero, intensified its scrutiny activity in 2012, especially when the provision of rescue loans to the Spanish banks was approved (Figure 7.8). This is consistent with previous research that illustrates that IU is the party that more often opposes EU-related legislation (Palau, Muñoz and Chaqués-Bonafont, 2014). Without turning euroskeptic, this party has increased its criticisms toward EU policies, especially regarding market deregulation, monetarist orthodoxy and cuts in the welfare state, and privatization of essential public services (Rodriguez Aguilera, 2014).

The PSOE also increased the percentage of oral questions devoted to these policies once in opposition. Similarly, regional parties brought their scrutiny activity on these policies to unprecedented levels, especially CiU, which had never been concerned before with EU economic and labor decisions. These questions however were not designed to openly criticize EU decisions but more to attack the incumbent; for example, urging it to adapt the Spanish Mortgages Act to the European Court of Justice Judgment or asking it how the economic recommendations of the European Commission to reduce unemployment rates would be implemented.

Conclusions

In this chapter we have illustrated that Spanish MPs devote their attention to EU concerns as a means to attack each other in the government-opposition game. Their goal is not to politicize European integration nor to question Spanish membership, but to use the EU frame in their competition with each other. While in government, parties give visibility to EU affairs depending on the opportunities provided by the institutional context – variations in agenda capacity resulting from reforms in the rules that govern parliamentary questioning – and the occurrence of EU events. They mainly use the EU to account for the results of EU meetings and negotiations and to give visibility to positive developments and successful policies. This was especially the case of the PP government, regarding the fulfillment of the convergence criteria and the successes achieved in the economic field.

Opposition groups use the EU frame to draw attention to negative events as they try to portray the government as a poor negotiator, unable to defend the national interest in Brussels, and forcing the executive to discuss issues it would rather avoid. This is especially the case of regional groups that use the EU to give visibility to regional interests and demands related to language issues or representation rights in front of European institutions. In line with the output-based form of legitimacy that has characterized public support for European integration, opposition groups also use critical junctures that occur when the proportion of citizens who believe that the country does not benefit from European integration increases to attack the government, especially regarding EU funds and agriculture issues and – since the outbreak of the crisis – for criticizing its handling of economic issues, using the EU frame to weaken it.

Overall, although important issues of public policy are moved upwards toward the EU, opposition groups do not use parliamentary question time to undertake routine parliamentary oversight on EU decisions nor to compensate for its secondary role in the EU policymaking process. They do not use the parliamentary arena to debate about the course of European integration but pay attention to EU issues when it makes partisan sense to do it, integrating Europe within the government-opposition game.

In the next chapter, we analyze the abortion debate, illustrating, among other things, that the international frame is still occasionally invoked by political parties to convince opposition groups and the public more generally to support their policy alternatives. Supranational ideas and developments are used in the construction of problems and ultimately to justify agenda setting and policy change.

8
Framing the Abortion Debate

Abortion is an issue where most citizens' views are solidly fixed, as they are related to a moral dimension and strongly associated with religious beliefs. These are highly stable because they are based on strong conceptions of right and wrong; people do not change their moral views on a whim, especially on an issue such as this where the Catholic Church plays an important role. Although abortion has predominantly been discussed in moral terms, the moral view is not the only way of framing about this topic. Multiple frames exist; abortion is a multifaceted issue. As with any issue, public attention may emphasize a particular dimension of the debate and tap into different frames. Thus, even on a moral issue with strong religious overtones, shifting foci of attention may allow some movement in aggregate public opinion and policy, at least on different elements of the debate.[1]

To highlight the importance of framing and shifting attention in Spanish politics, we analyze the Spanish abortion debate in this chapter. The goal is to trace the shifting foci of attention to abortion in parliamentary debates and on the front pages of the two leading newspapers in Spain, *El País* and *El Mundo*, to answer a series of question about the interplay between frames, political parties, and the media. Do parties or coalitions in Parliament shift attention from one aspect of the issue to another over time to promote or avoid policy change? Do the media emphasize certain aspects of the debate by giving support to the framing strategies of particular political parties and entrepreneurs while ignoring others? Is media coverage predominantly positive, negative, or neutral with regard to abortion? Do trends in the tone of media coverage relate to shifts in attention in Parliament? To do this systematically, we constructed a list of seven distinct frames – morality, fairness, efficacy, education, constitutionality, crime and law, and international compatibility – used for or against abortion. Then we analyzed the tone – positive, negative, or neutral – of media stories that tapped these frames.

Results illustrate that Spanish parliamentary groups emphasize different dimensions of the abortion debate to support their political strategies and preferences. Policy change and stability, however, cannot be attributed only to framing and changing preferences as opposed to other factors. Changing religious values and institutional factors, in particular the control of government by left or right parties and the type of government, are important predictors of agenda access and policy change. That is, political control of government by left or right leads to policy change in this case – and this is done straightforwardly, not by a rhetorical flourish unanswered by the opposition. The minority governments of the 1990s blocked all attempts of reform made by left parties for broadening the abortion regulation of the 1980s. Similarly, the conservatives would not have attempted to pass a radical and controversial reform on abortion without enjoying an absolute majority of seats in the Chamber.

In this chapter we also show that the media is a major source of instability in Spanish politics. The analysis of media stories in *El País* and *El Mundo* illustrates that both newspapers strongly react to the parliamentary initiatives introduced in the Chamber by parliamentary groups and use different tones, emphasize different dimensions of abortion as a problem, and give visibility to distinct sets of entrepreneurs in order to promote a particular aspect of the debate. The framing of abortion in *El País* is closer to the framing strategies of the Far Left (IU) and the PSOE, and conservative dimensions of the debate are highlighted in *El Mundo*, which more closely resembles issue frames used by the PP. Overall, the media contributes to positive-feedback dynamics, which reinforce punctuated equilibrium tendencies that exist in other areas of the policy-making (Baumgartner and Jones 1993). Before turning to the analysis, we first provide an overview of the abortion debate in Spain.

The abortion debate

Political parties in Spain have not tried to change basic citizen beliefs about abortion; rather, they have focused on different attributes of the issue in order to promote (or avoid) policy change. In the early 1980s, the PSOE highlighted the *efficacy* of abortion to avoid major problems (i.e., protecting a potential mother's physical or mental health) and the *crime and law* perspective (i.e., highlighting that a regulatory reform was necessary to avoid women and doctors going to prison for practicing abortions), in order to pass a law in 1985 that decriminalized abortion under some circumstances. The conservatives, in contrast, framed the issue in *moral* terms, emphasizing the unborn fetus' right to life, and in *constitutional* terms by sending the law to court arguing that it violated Article 15 of the Constitution recognizing the rights to life and to physical and moral integrity.

During 1990s, left parties, with the support of feminist movements, kept abortion on the agenda following a strategy of issue expansion based on criticism (Baumgartner and Jones, 1993). They politicized the issue by not only referring to a *moral* perspective – a woman's right to choose – but also to the *inequity* caused by the existing regulatory framework. Additionally, they pointed out the *legal uncertainty* that most women and doctors faced at the time resulting from the comprehensive interpretation of some of the exceptions defined by the 1985 regulation. The *international* dimension was also introduced in the debate arguing that a regulatory reform was necessary so that Spanish legislation could be brought in line with that of other Western democracies.

Both the PSOE and IU generally emphasized these dimensions of the problem, but each made them their own. The Far Left (IU) continuously kept abortion high on its agenda. They defended the adoption of the socialist legislation and urged its extension toward more liberal positions, recognizing women's right to decide under any circumstance, from the beginning of the debate. In contrast, the PSOE was initially concerned about the implementation of the 1985 law; especially regarding the asymmetries caused by the moral objection of doctors in terms of access and the low number of abortions practiced in public health institutions. Until the mid-1990s, the PSOE did not actively ask and campaign for the passage of a new regulatory framework. During the 1990s, all the attempts made by left parties to pass a new bill failed because regional conservative parties, mainly CiU and PNV, and the PP, vetoed all the parliamentary initiatives introduced by the PSOE and IU to reform the abortion law during the minority governments of González and Aznar.

Despite adopting policies close to the positions of the Spanish church in policy areas like education, the conservatives followed a reactive strategy regarding abortion and only intervened in parliamentary debates when liberalization appeared imminent. The PP did not try to change existing regulation after winning the 1996 general elections, when it governed as a minority government, nor when it had an absolute majority of seats in the parliament from 2000 to 2004. Prime Minister Aznar (PP) did not push to change extant abortion policy, despite pressure from the

Catholic Church to do so (Magone, 2004: 292). Beginning in 1989 when ideological moderation characterized Spanish governments generally, the abortion debate was largely silenced, falling off the conservative agenda, until the socialist government of José Luis Rodríguez Zapatero adopted an important reform in 2010.

The PSOE government introduced a "period model" that recognized the reproductive rights of women in their 2010 reform. Despite their minority government status, the socialists garnered the parliamentary support to pass this new law and other reforms concerning morality issues, like the legalization of same-sex marriage (2005) and new regulations concerning assisted reproductive technologies (ART), by appealing to their regional allies (the ERC, the BNG, and part of the PNV). The abortion reform was passed emphasizing different dimensions of abortion as a problem, especially the *unfairness* of the existing regulatory approach, an aspect of the debate that had been highlighted by left parties in the Spanish Parliament for more than a decade. The PP, pro-life associations, and the Catholic Church strongly criticized the reforms. The conservatives sent the abortion law to the Constitutional Court for constitutional revision and highlighted *constitutional* and *moral* arguments, such as the fetus' right to life. Additionally, they introduced a new *moral* argument into the debate: therapeutic abortion is inadmissible because differentiating between healthy and handicapped fetuses is an unacceptable form of discrimination.

After reaching power in 2011, the PP took an active position to reform the abortion law for the first time since the late 1980s. Increased power of the pro-life faction within the PP, led by new Ministry of Justice, Alberto Ruíz Gallardón, and pressures from the Church and pro-life associations led the conservatives to introduce a bill that proposed more restrictive regulations than the initial bill passed under the socialist government in 1985. This reform proposed to guarantee abortion only under two circumstances: in case of rape or when the mother's health is seriously at risk, stripping away the provision that allowed abortions in case of the fetus having a serious disease (with the exception of those fetuses incompatible with life). On the basis of *constitutional* and new *moral* arguments related to the unacceptable discrimination of handicapped fetuses, the Conservatives brought abortion to the forefront of the political agenda on the grounds of strong confrontation and polarization. In this sense, the abortion is a clear example of the new way of doing politics characterized by the increasing lack of consensus that we have been describing over many chapters in this book.

Expanding on this overview of abortion policy and associated frames, this chapter is organized as following. We first explain and expound on the multiple dimensions of abortion as a problem using the methodology developed by Baumgartner et al. (2008) for measuring and analyzing framing and issue definition over time. Next, we analyze parliamentary attention to abortion, illustrating the framing strategies followed by left and right parties. Finally, we focus on media attention and how the dimensions highlighted by the two major newspapers in Spain parallel that of the political parties.

The multiple dimensions of abortion

Our analysis of abortion is based on seven main dimensions of the debate: morality, fairness, efficacy, education, constitutionality, crime and law, and international compatibility (see Table 8.1). Both sides have utilized *moral frames* in the abortion debate. In fact, these frames provide the labels ascribed to the two distinct sides in many portions of the English-speaking world: pro-life and pro-choice. Those that defend the pro-choice option argue that women have the right to decide when abortion is morally justified and whether they want to continue or interrupt the pregnancy. On the other hand, those against abortion tap right-to-life arguments focusing not on the mother but on the unborn fetus and emphasize the need to protect and respect life from the moment of conception. They also emphasize the father's right to decide about the pregnancy as well as the objections of doctors who do not want to practice abortions. Not only is abortion morally unacceptable to those on the pro-life side of the debate, but the use of contraceptives that prevent pregnancies is also considered morally unacceptable. These moral arguments may or may not be related to religious arguments concerning the "will of God." However, due to the increasing secularization of Spanish society (Requena, 2005), religious aspects of the moral arguments have been increasingly left out of the debate.

Abortion is also discussed in terms of *fairness*. Pro-choice groups argue that a restrictive regulation where abortion is not free and funded by public institutions is unfair because only women with the requisite economic resources have access to safe abortions, for example, by traveling abroad. From this perspective, a liberal regulation and a system of public health institutions that guarantee access to abortion is considered necessary to avoid public health problems such as the woman's death due to clandestine abortion practices, or inequalities in terms of access resulting from the moral objection of doctors that may result in

Table 8.1 Dimensions of the abortion debate

Dimensions	Pro-abortion arguments	Anti-abortion arguments
Morality	Women's rights to choose whether they continue or interrupt pregnancy, and to decide when abortion is morally justified The abortion pill and contraceptive measures are necessary to guarantee woman's rights regarding maternity	Abortion is against the "will of God" Unborn right to life: human life must be respected and protected from the moment of conception The abortion of a fetus with anomalies or handicapped is an unacceptable form of discrimination Father's right to decide about the pregnancy Moral objection of doctors to practice abortions The use of contraceptive measures are morally unacceptable Both the abortion pill and the "morning-after pill" are abortion mechanisms: chemical weapons designed to kill innocent babies
Fairness	Denial of abortion right is unfair because only those women with resources may have access to a safe abortion. The rest are forced to have clandestine abortions with high risks for women's health. Right to have access to a safe abortion financed by public institutions All arguments related to the difficulties or lack of access to the public system All pro-abortion arguments that stress socioeconomic aspects and inequalities (e.g., suffering immigrants and economically disadvantaged women)	Abortion represents a physical and psychological aggression for women, an unfair option chosen only by those women that have no resources to go ahead with the pregnancy The role of public institutions is not to facilitate abortion but to give women without resources the financial and moral support to raise their children Women may be pressured or forced against their will to have an abortion (by their partners, social environment, etc.)
Efficacy	Abortion is effective because it prevents major problems It protects mother's life, avoids useless suffering in cases where the unborn child has an incurable or serious disease or in cases of sexual assault	Abortion is ineffective because it promotes the use of abortion as a contraceptive mechanism, increasing the overall number of abortions Contraceptive mechanisms are effective and necessary to avoid abortions

	The abortion pill is effective because: (1) it avoids the suffering and trauma associated with a clinical abortion; (2) reduces medical and surgery expenses; and (3) avoids medical and health risks Contraceptive mechanisms and the "morning-after pill" are acceptable and effective measures to avoid unwanted pregnancies A law based on time limits is effective to avoid late abortions allowed by the 1985 law	The "morning-after pill" is not an abortion mechanism and is effective to avoid unwanted pregnancies
Education	Sexual education to avoid unwanted pregnancies must complement policies that recognize women's abortion rights	Abortion is never justifiable; sexual (and/or religious) education is necessary to avoid unwanted pregnancies
Constitutionality	Abortion is not against the Constitution. The 1985 Constitutional Court sentence recognizes the intrinsic value of the fetus but denies his right to life	Abortion is against the Constitution that recognizes the right to live The Constitutional Court sentence does not recognize the unlimited right of women to decide
Crime and law	Women and doctors practicing abortion should not go to prison under any circumstance All damages and attacks to abortion defenders and clinics must be punished by law A legislation based on exceptions (and not on a time period) leaves women and doctors in a highly insecure legal position, transferring the burden of proof to doctors and promoting the criminalization of abortion	Women and doctors practicing abortion should go to prison, as they are committing a crime punishable by law Only doctors (not women) practicing abortions should go to prison as they are ultimately responsible for the action The "morning-after pill" is against the 1985 law, an abortion mechanism used when none of the three exceptions considered by the legislation are present All arguments related to the adequacy of abortion practice to the applicable legislation (compliance)
International	A more liberal regulation of abortion is necessary to equate Spanish regulation to that of other Western democracies	The right to life is recognized in international agreements ratified by Spain No international treaty defines abortion as a right

territorial asymmetries. Pro-life advocates, in contrast, use this frame to argue that abortion constitutes a physical and psychological aggression toward women and represents a harsh and unfair measure because it is an option that women, and especially the poor, choose simply because they do not have the sufficient resources to go ahead with the pregnancy, or because they are under unacceptable pressure from their partner or the social environment to have an abortion. The role of public institutions therefore is not funding abortions but giving these women resources for raising their children, according to this view.

Those against abortion question its *efficacy*, because they claim a more liberal regulation would lead to its increased use as a contraceptive measure, resulting in a large increase in the number of abortions. On the other hand, supporters of legislative liberalization argue that abortion is an effective method of avoiding major problems such as in the cases of useless suffering of the unborn child if s/he had an incurable disease. The abortion pill is considered effective because it avoids the suffering and trauma associated with a clinical abortion; it reduces medical and surgery expenses; and it avoids medical and health risks that are associated with clinical abortion. Those that do not consider the "morning-after pill" as an abortion mechanism, per the definition of the World Health Organization (WHO), also use efficacy-related arguments to defend its effectiveness to avoid unwanted pregnancies and ultimately avoid abortions. Contraceptives and the morning-after pill are also considered effective measures to avoid unwanted pregnancies. In sum, the efficacy frame focuses on what reduces the need for abortion or reducing suffering.

Within the *education* frame, those that oppose abortion argue that to put an end to a human life is never justifiable and that (religious) education is the best way to avoid unwanted pregnancies. Pro-choice advocates, in contrast, consider sexual education necessary and must complement the recognition of a woman's right to choose. They argue that increased access to sexual education is necessary to prevent more abortions.

Another dimension of the abortion debate relates to the *constitutionality* of regulations. The Constitutional Court has the right to review legislation and can totally or partially invalidate laws on the grounds that they violate articles of the Constitution, including the right to life. At present, judicial review in Spain can only take place a priori when it involves the ratification of international treaties. However, until 1985, a priori constitutional censure – judicial review following parliamentary approval of a piece of legislation but before it was implemented – could take place for organic laws (Barreiro, 1998; 2000). This frame was used in the 1980s by parties opposed to the liberalization of abortion laws to delay the implementation of reforms. Following the elimination of this a priori judicial review, constitutional revision is still being used by opponents of policy change to obtain later revocation of legislation, and to signal their opposition to their voters.

Because facilities and physicians providing abortions must do so within the framework of the 1985 law and any amendments thereafter, the debate about abortion involves a *criminal and legal* issue frame. This dimension has been used by left parties to support a law based on time limits (period model) to highlight the legal insecurity faced by women and doctors that practice abortions under the legal exceptions, especially that of psychological harm for the mother, which is the most often cited rationale to end a pregnancy. In contrast, pro-life advocates use this dimension to argue that all abortions practiced outside the scope of the 1985 legislation must be punished. Some with more radical positions argue that both the woman and her doctor should go to prison, while others defend only the criminalization of doctors, as the ultimate responsibility for abortion practices lies with them. This argument has been increasingly used by pro-life groups to counter the criticism of being anti-woman, a charge that they often faced from pro-choice advocates.

Finally, the *international* dimension is an important aspect of the debate. Most Western democracies have a "period model" law that allows women to terminate pregnancies within the first 14 weeks without having to give special reasons. Pro-choice advocates point to this fact and argue that by adhering to the status quo Spain is moving away from what most countries have done for the recognition of women's rights. On the other hand, those that are against the liberalization of abortion regulations argue that international human rights charters and treaties recognize the right of all to life and do not recognize abortion as a right; therefore, a liberal regulation on abortion is against international agreements ratified by Spanish authorities.

The definition of these seven dimensions allows us to trace which frames are dominant at various points in time and also facilitate the analysis of how political actors, from political parties to the media, address the issue. To analyze the arguments highlighted by political parties we have considered all the parliamentary initiatives (executive and parliamentary bills, oral and written questions, and interpellations) introduced in the Spanish Parliament from 1982 to 2011. For the study of the media-framing strategies, we have analyzed all the stories published on the front pages of *El Mundo* and *El País* from 1996 to 2011.

Each story and parliamentary question was coded multiple times if it mentioned multiple frames.

In the case of media stories, each frame was further divided into positive, negative, and neutral arguments. We considered a story to have a positive tone when it reported facts or data that would be positively evaluated (good news) for those defending a pro-choice stance (e.g., "the Spanish law recognizes for the first time women's right to abortion"). We considered a story to have a negative tone when it promoted a view or described an event leaning in a pro-life or anti-abortion direction (e.g., "massive protest against the new abortion law passed by Zapatero"). We considered a story to have a neutral tone when both those defending a pro-choice and a pro-life stance could agree on the positive or negative tone of a specific story (e.g., "the number of abortions have increased among teenagers"). This could be a negative story for both groups if we consider that by definition abortion is a bad thing. Other neutral stories focused on equal parts pro-choice and pro-life frames; they were equally balanced or simply reported a fact that was not clearly interpretable either way. We start by explaining how parliamentary groups frame the abortion debate in the Parliament.

Framing strategies in Parliament

Public policies are not always ruled by incrementalism; rather, there are critical periods of mobilization and agenda access during which dramatic changes may occur (Baumgartner and Jones, 1993). In the regulation of abortion, the first important reform after Francoism was introduced by the Organic Law 9/1985 passed by the socialist government. This reformed one article of the penal code that kept the definition of abortion as a crime punishable by law, but recognized three exceptions under which a woman can terminate her pregnancy: if pregnancy is the result of rape or an act of incest that has been reported to the police; if the fetus is severely physically or mentally handicapped; or to avoid physical or mental harm to the mother.[2] As Figure 8.1 illustrates, this reform set off a first wave of attention to abortion in the Spanish Parliament that was led by the conservative party (PP) that strongly reacted to the socialist reforms.

The 1985 law was the result of a bill introduced by the PSOE in 1983 fulfilling its electoral promise[3] of reforming abortion regulation after the transition to democracy. To promote policy change in 1985, the Socialists framed the issue in terms of *efficacy*, underlining the convenience of legalizing therapeutic and ethical abortions to prevent major problems,

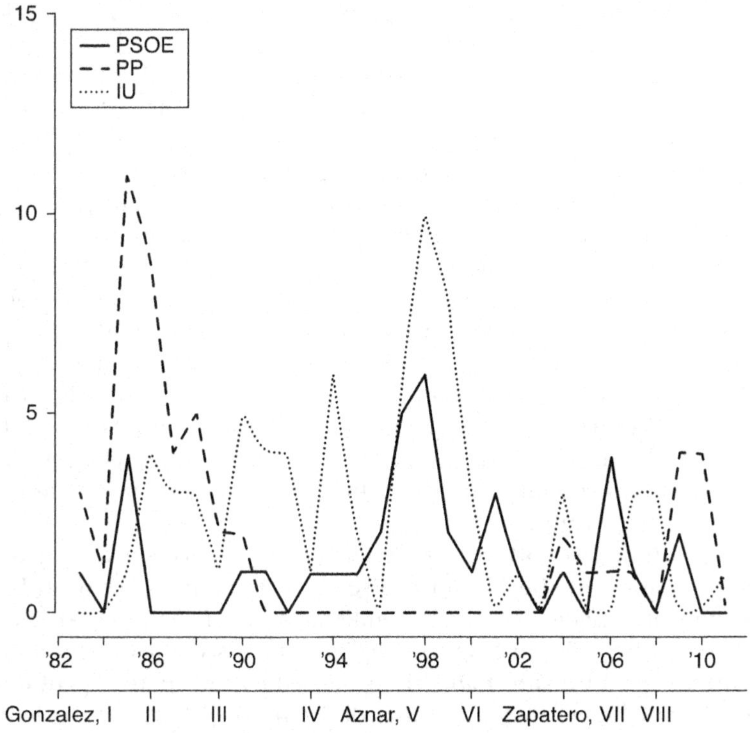

Figure 8.1 Number of parliamentary initiatives introduced on abortion

Note: The figure shows the number of parliamentary initiatives (oral and written questions, executive and parliamentary bills, and interpellations) introduced for each political party shown, on the topic of abortion. Data are annual.

Source: Spanish Policy Agendas Databases (www.ub.edu/spanishpolicyagendas).

as Figure 8.2 illustrates. Additionally, they emphasized the *crime* frame by highlighting that a legal reform was necessary to avoid women and doctors practicing abortions going to prison. The 1983 bill was framed according to these two dimensions leaving the moral dimensions in the background. The debate was not centered on whether women should have the right to decide when to terminate pregnancy but on whether a more liberal regulation was necessary to protect a mother's life and to avoid useless suffering in case the unborn child had an incurable disease by decriminalizing abortion under some circumstances. The PSOE did not emphasize a moral perspective until 1985. It did not do so to openly defend women's right to decide under any circumstance, but to give visibility to the implementation problems caused by the moral objections of doctors.

Members of conservative parties, like UDC or PNV, gave also support to the *efficacy* and *crime* framing, and only the PP and the Catholic Church were completely against the recognition of the right to legal abortions in a few concrete situations, except in the case of a threat to mother's life, a circumstance already accepted as standard practice in the Francoist Penal Code (Barreiro, 1998: 153). Contrary to the PSOE, the PP mainly framed the issue in *moral* terms following the principles and values of the Catholic Church (Figure 8.2.a). The conservatives highlighted in parliamentary initiatives the fetus' right to life and tapped other moral arguments, without explicitly mentioning religion. Additionally, the *constitutional* dimension played a major role in the Conservatives' framing strategy. Once the socialist bill was passed the PP referred it to the Constitutional Court, questioning, among other things, its compatibility with the right to life recognized in Article 15 of the Spanish Constitution.

At this time, a priori judicial review became a weapon exercised by the opposition in the face of the socialist absolute majority to both block regulatory change and to limit the scope of the political reform (Barreiro, 1998). Eleven days after the bill was passed, the Constitutional Court accepted the direct appeal introduced by the PP, and spent more than a year before making a final decision. The Court ruling denied both the absolute right to life for the fetus and the woman's freedom of choice and asked for changes in the bill designed to provide greater protection of the fetus and to regulate the actual carrying out of abortion. The changes asked for by the Court were: before preforming an abortion, doctors were asked to assess the potential consequences of the pregnancy on the physical and mental health of the woman; responsibility should be placed on the medical profession to see that at least one of the three conditions under which abortion is legal was met; and abortions had to be carried out in public health institutions or private clinics authorized by the state. The final bill included amendments necessary to satisfy the Court ruling, which was finally passed in 1985.

Following the passage of the first reform law in the 1980s, the possibility to liberalize the abortion law were limited because of the existence of a priori judicial review. Additionally, the possibilities were limited because after 40 years of Francoism and cultural domination by the Catholic Church, public opinion favored the legalization of therapeutic, eugenic, and ethical abortions (63 percent of citizens according to a CIS poll conducted in 1983) but did not support further liberalization. As

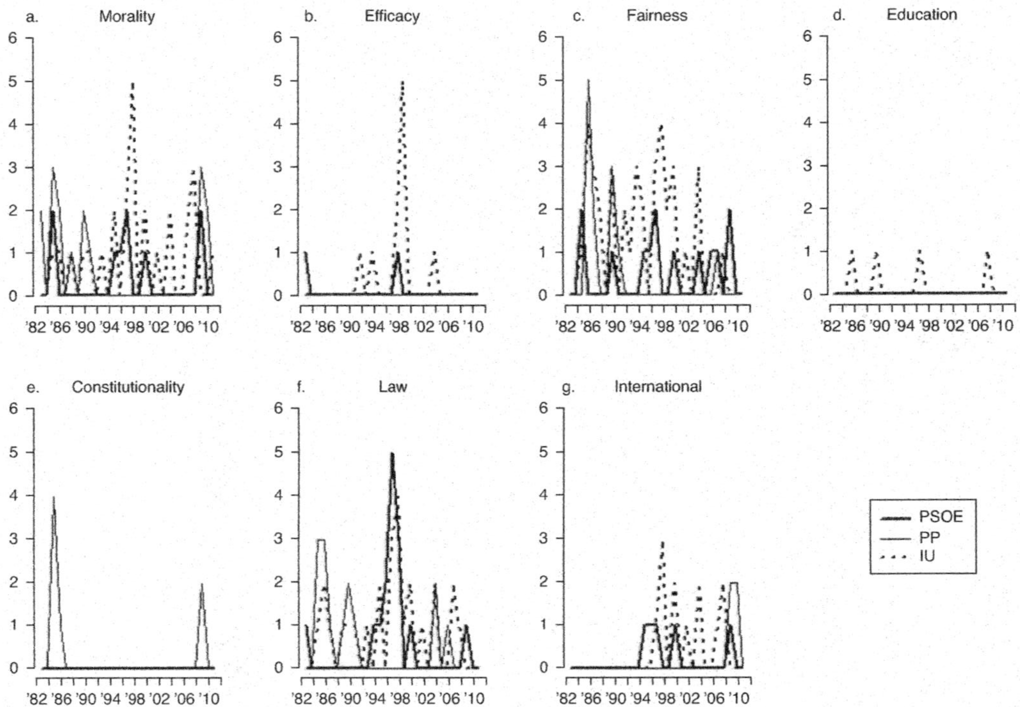

Figure 8.2 Frame use over time by political party

Note: The figure shows the number of parliamentary initiatives (oral and written questions, executive and parliamentary bills, and interpellations) that use each of the seven frames indicated, for each of the three parties indicated. Data are annual.

Source: Spanish Policy Agendas Databases (www.ub.edu/spanishpolicyagendas).

Figure 8.3 illustrates, on a 1 to 10 scale (1 abortion is never justifiable and 10 it is always), the mean for Spanish citizens was less than 3.0 in 1981 (54 percent saying that it was never justifiable). The 1985 law was a conservative model of regulation that represented a first step toward liberalization that fit the Court requirements and public opinion mood but did not respond to either the policy preferences of the Far Left, whose aim was the total decriminalization of abortion, or the Right, whose aim was to keep abortion as a crime. As a result, following the Court ruling and the passage of the Organic Law 9/1985 there was an important change in the framing strategies of both Left and Right parties.

While the conservatives kept emphasizing *moral* principles (Figure 8.2.a), they also introduced new dimensions to the debate such as the

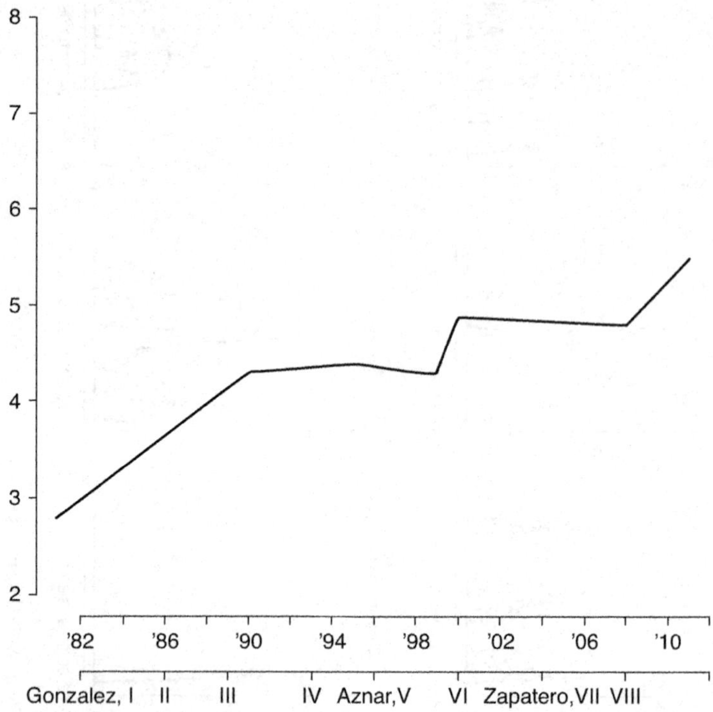

Figure 8.3 Public opinion on abortion

Note: The figure shows the mean response from national surveys of the Spanish adult population on a scale running from 1 ("abortion is never justifiable") to 10 ("abortion is always justifiable"). Data are annual.

Source: World Values Survey (www.worldvaluessurvey.org).

crime and law and *fairness* perspectives. The PP renounced reform of the abortion legislation as a goal and instead focused its efforts on guaranteeing the compliance of abortion practices with the existing regulatory framework. The conservatives introduced many questions focused on details of implementation, avoiding any comprehensive attack on the law as a whole, but focused especially on the issue of the doctor's assessment of possible psychological harm for the mother, as most abortions were being practiced under this provision of the 1985 law.

Conservative members of Parliament also emphasized the *fairness* perspective by asking the socialists questions related to whether they had implemented policy measures designed to provide women who wanted to carry out their pregnancy with economic support to avoid electing to have an abortion out of economic necessity. The emphasis given to these dimensions is also illustrated in parliamentary question sessions but also in a bill introduced by the PP in 1986 (and again in 1989) designed to create a Commission on the Defense of Life (*Comisión en Defensa de la Vida*) whose functions would be to monitor the implementation of the 1985 law and to develop policy programs in support of women's economic support to carry pregnancy to term.

In the late 1980s, the conservatives decided to leave abortion off of the agenda in the framework of a catch-all electoral strategy and programmatic moderation designed to gain power (Gunther and Montero, 2009). After winning the 1996 general elections, the PP did not consider reforming the legislation passed by the Socialists in the 1980s, neither did it do so when it gained an absolute majority in 2000. As Figure 8.3 illustrates, after introducing a significant number of initiatives during the mid- and late 1980s, the PP followed a reactive position and never committed to reform the Socialist legislation. Although José María Aznar implemented policies close to those on the agenda of the Catholic Church, like the restoration of obligatory religious instruction in schools or the limitation of ART (assisted reproductive techniques) and SCR (stem-cell research), it did not reform the abortion law in spite of pressures from the Church hierarchy to do so (Magone, 2004: 292). This unwillingness by the PP government to revise the 1985 law relates to a strong public consensus around the regulatory framework defined by the 1985 law.

In the case of Left parties, because the abortion law did not satisfy their expectations, they used the parliamentary arena as a venue to keep the issue alive. As Figure 8.1 shows, this was especially true of the Far Left (IU), who constantly pushed abortion onto the parliamentary agenda to promote a more liberal regulation after 1985. To do so, it followed an overall strategy of issue expansion and mobilization through criticism based on politicizing the abortion debate through criticism of existing arrangements and institutions (Baumgartner and Jones, 1993; Schattschneider, 1975). The parliamentary initiatives introduced by the IU focused on different dimensions of abortion as a problem emphasizing disagreement, fear, mistrust, and complaints to promote and legitimate a new regulatory reform.

The Far Left used the *moral* frame to highlight women's right to decide about pregnancy and gave special visibility to how the moral objection of doctors made access to abortion difficult, especially in some regions. The *fairness* perspective was used to emphasize the difficulties to get access to the public system, as most of the abortions were being practiced in private clinics and the percentage of those financed by public institutions was very low. The *crime and law* dimension was used to highlight the pitfalls of the regulatory system in terms of the situation of legal uncertainty that the 1985 law created for doctors, who had the ultimate responsibility for authorizing abortions and could assume legal charges if the exceptions set out by legislation were not appropriately justified.

The PSOE also followed a strategy based on criticizing aspects of the 1985 legislation, but contrary to the IU it did not initially advocate for reform of the regulatory framework. Until the mid-1990s, members of the PSOE paid less attention to abortion than the Far Left (see Figure 8.1). Once the 1985 law was passed, the socialists were concerned about implementation issues and, like the Far Left, highlighted in parliamentary oversight sessions the doctor's moral objection problem, the difficulties associated with access to abortion through the public system, and the criminalization of doctors. However, it did not introduce any bills designed to introduce a regulatory reform until the mid-1990s. In its 1989 party manifesto, the socialists only mentioned the party's commitment to a more efficient implementation of the law, in contrast to the IU, which was already calling for a law based on time limits. In 1993, the party manifesto of the socialists was more ambiguous, mentioning that women's decision to finish a pregnancy should be a free decision, but it did not openly commit to any reform of the 1985 legislation.

Following the 1993 general elections, when the socialists lost the absolute majority of seats in the Chamber, the legislative alliance with the Catalan Conservative regional party to pass legislation compromised any reform of the abortion law toward more liberal positions. The socialists only introduced an executive bill[4] in 1995 at the end of the legislature when the government of Felipe González had already run out of steam due to several political scandals and had almost lost the support of CiU. The call for early elections when the government failed to get the budget bill approved in December 1995 eliminated any possibility of having this bill passed. It was mainly used to satisfy its more Left-oriented electorate.

In the 1996 elections campaign, the PSOE committed to pass new legislation reforming abortion regulations, but it lost the election, which meant that institutional factors limited any possibility to pass a more liberal regulation. The PP, which opposed any type of reform, was governing with the support of the conservative Catalan and Basque parties (CiU and PNV). This resulted in a conservative veto to all bills introduced by IU and the PSOE, which as Figure 8.1 shows led a new wave of attention to abortion in the Spanish Parliament. The fact that there was no chance for passage of new abortion legislation during this period, but these parties introduced numerous bills to do exactly that, shows an important aspect of the use of parliamentary opposition tactics in order to raise issues making life difficult for the government. But it was not purely opportunistic, as the parties followed through with some reforms when they eventually came to power, as we discuss below.

The PSOE was much more active compared to previous legislatures and introduced a significant number of initiatives, including a bill that proposed the inclusion of a fourth circumstance under which women would be allowed to practice abortion: when the pregnancy represented a personal, family or social conflict. The possibility of decriminalizing a new circumstance explains way the debate was mainly framed in terms of *crime and law* (Figure 8.2.f), but the *fairness* and the *moral* dimensions were also important elements of the Left-framing strategies, especially for IU. As Figure 8.2.g shows, during this period a new frame related to the *international* dimension also emerged. Both the PSOE and IU introduced for the first time international arguments to justify a new regulation of abortion presenting reforms as necessary to bring Spanish legislation into line with that of other Western democracies. The *efficacy* argument also peaked at that time (Figure 8.2.b) as a result of the abortion pill debate that occurred in 1999 when the PP agreed on the introduction of the abortion pill, a measure recognized by the WHO, that had already been authorized in most EU countries. Left parties strongly pushed the Conservatives toward authorizing this pill, highlighting its efficacy in terms of avoiding the suffering, trauma, and surgery costs associated with clinical abortions.

The arrival of Zapatero into power allowed for a major regulatory reform on the basis of pro-choice arguments that had been highlighted for more than a decade by Left parties in the Spanish Parliament. Zapatero prioritized abortion in the governmental agenda to fulfill campaign promises outlined in the 2004 party manifesto within a broader agenda of reforms regarding other morality issues. In 2005, the socialists legalized same-sex marriage. In 2006, they passed a new regulation concerning ART and SCR that allowed savior siblings to be born ("savior siblings" are younger siblings conceived by their parents because an older child has a medical condition that can only be treated with a transplant or transfusion from a sibling). In 2010, the socialists and their allies reformed the abortion law, substituting a fairly liberal regulatory model in place of the restrictive 1985 law. The Organic Law on Sexual and Reproductive Health and on the Voluntary Interruption of Pregnancy (Law 2/2010) put an end to the criminalization of abortion and introduced a "period model" that recognized the reproductive rights of women. According to this law, women have unlimited individual choice within the first 14 weeks of pregnancy to decide whether to terminate pregnancy without having to give special reasons and without requiring medical authorization. To pass this legislation, the PSOE bargained with other political parties to build a coalition that ensured the passage of the regulatory changes. Despite the socialists forming a minority government, they had the support of the Far Left (IU), and regional political parties like ERC, BNG, and part of the PNV, to pass the new abortion law, a situation which was radically different than that of the mid-1990s.

The socialist agenda of morality reforms involved the reemergence of abortion in the conservative agenda following almost two decades of passive and reactive position taking. As Figure 8.1 illustrates, conservative parties like the PP and UPN strongly reacted to the socialists' plans by introducing a large number of initiatives in the Chamber that led to the politicization of abortion. Conservative parties and the Church hierarchy, with the support of increasingly powerful pro-life associations, manifested a strong opposition to the abortion reform as well as to other decisions adopted by the socialists, like the downgrading of religion as part of the school curriculum and the reduction in state financial support for the Church (Magone, 2004). As a result, the Episcopal Conference openly criticized the Zapatero government and called for religious citizens to vote accordingly, ending more than 20 years of political neutrality of the Church, and reviving a religious cleavage that had been dormant for nearly three decades (Gunther and Montero, 2009: 244).

As in the 1980s, the conservatives framed the reform by appealing to the *constitutional* frame. The PP sent the bill to Court for constitutional revision but as the possibility of prior judicial review was abolished in 1985 it could not use it as a means of parliamentary obstruction and implementation delay. However, challenging the law in Court was still a good opportunity for providing visibility to the party's positions toward moral issues. Additionally, it signaled to their religious members and the Catholic Church that the party had not abandoned the fight. The *international* dimension was emphasized also by the Conservatives at that time to argue that the socialist legislation was against international charters of human rights that had been ratified by Spanish authorities (Figure 8.2.g).

The PP also used the *moral* frame by recycling old arguments, like the fetus' right to life and the conscientious objection right of the medical profession. Additionally, they highlighted a totally new dimension of the moral debate; they argued that the Zapatero law was unconstitutional because all human beings, even those seriously handicapped or with serious diseases, have the right of not being discriminated against. To practice an abortion on the grounds of disability or sickness is a negative form of discrimination as inadmissible as those based on race or gender. This position was adopted in line with the arguments defended by the CERMI (*Comité Español de Representantes de Personas con Discapacidad*), a platform that brings together a number of organizations defending the rights of handicapped people in Spain.

This represents a strong departure from previous positions and a serious conservative turn in the party position regarding abortion. This turn must be understood in the broader framework of polarization that currently characterizes Spanish politics (see Chapter 3). This departure became clear after the PP won the 2011 general elections with an absolute majority when the government of Rajoy, under the initiative of the strongly conservative Minister of Justice Alberto Ruíz Gallardón presented an abortion reform bill that included the possibility of abortion only under two suppositions: when the mother's health was seriously at risk or in the case of rape. The proposed bill stripped away the provision that allowed for abortions in case of the fetus having a serious disease with the only exception being those fetuses incompatible with life. This reform reopened the abortion debate in the parliamentary arena and generated a cleavage within the conservative party, which has strong supporters for the reform but also opponents among the more moderate members of the party. Additionally, this bill was opposed by the medical profession, which strongly criticized the lack of inclusion of the serious fetal disease exception. Although the reform was supported by the Catholic Church and pro-life associations, including those defending the rights of handicapped people, the lack of support among public opinion, political parties from the opposition, and the medical profession forced Prime Minister Mariano Rajoy to announce in September 2014 the withdrawal of the bill after three years of debate. This caused the resignation of Ministry of Justice Alberto Ruíz Gallardón and strong criticism from both Right and Left parties that consider this initiative as one of the major failures of the legislature. The PP may still reform the abortion regulation passed by Zapatero but include less far-reaching reforms like the prohibition for minors to have abortion without parental permission.

We can see from the foregoing discussion that various political parties and governmental leaders have presented a great number of diverse arguments with regard to the abortion debate and have promoted or avoided the issue at different times in recent political history. Far from being a set issue, the theme of abortion comes and goes in the Spanish political agenda and is framed sometimes in moral, legal, and other terms. In the following section, we analyze to what extent the media contribute to the politicization of morality issues. Additionally, by emphasizing certain aspects of the debate, we analyze to what extent they give support to the framing strategies developed by Left and Right parties in the parliamentary arena.

Framing and media attention to abortion

As we have illustrated in Chapter 5, the Spanish media system is characterized by strong political parallelism. The two most widely read newspapers, *El País* and *El Mundo,* have substantial differences in terms of Left-Right orientations and strong connections with the PSOE and the PP, respectively (Chaqués-Bonafont and Baumgartner, 2013; Baumgartner and Chaqués-Bonafont, 2015). As a result, we expect that the different strategies followed in the parliamentary arena by political parties in the abortion debate translate into the media agenda so that the two newspapers emphasize different dimensions of abortion as a political issue. The analysis covers the period 1996–2011, as the media databases have been available since the mid-1990s.

Results illustrate that overall *El País* pays more attention to abortion than *El Mundo*. From 1996 to 2011, *El País* published 198 stories on abortion while *El Mundo* published only 86. In terms of the total media agenda, these represent 0.4 percent and 0.2 percent, respectively, of the total stories published in the front pages of these two newspapers. As Figure 8.4 shows, attention peaks in both newspapers when the

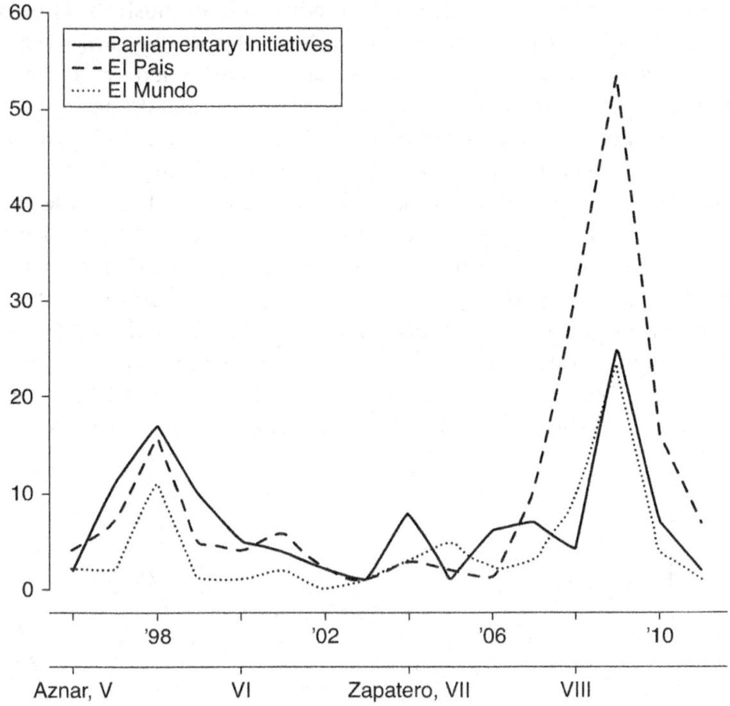

Figure 8.4 Parliamentary initiatives and news coverage of abortion

Note: The figure shows the number of parliamentary initiatives (oral and written questions, executive and parliamentary bills and interpellations) and the number of news stories in the front pages of the two newspapers indicated on the topic of abortion. Data are annual.

Source: Spanish Policy Agendas Databases (www.ub.edu/spanishpolicyagendas).

possibility of a regulatory reform enters the parliamentary agenda in 1998 following the parliamentary bills introduced by Left parties under the minority conservative government of the PP. They peak again in 2009 once the socialists introduce the executive bill to reform the 1985 legislation. This bill captured much more media attention than the failed attempt of the late 1990s, especially in the front pages of *El País*. The increase in news stories is mainly due to the politicization of the issue and the high probability of the passage of the reform. The annual correlation between the number of parliamentary initiatives and the number of media stories on abortion is positive and statistically significant in both newspapers (0.740 in the case of *El País* and 0.805 in the case of *El Mundo*, and both are statistically significant at the .05 level). It seems that in the case of abortion the media does not push the issue into the agenda to call the attention of Spanish MPs, but it mainly reacts to the initiatives taken by political parties in the parliamentary arena.

If we turn to the tone of media stories, results illustrate no significant differences between the newspapers. *El Mundo* publishes 47 percent of stories with a negative tone, namely that emphasize aspects that would be negatively evaluated for those defending the pro-choice side of the debate, and *El País* 51 percent. *El País* gives also slightly more visibility to stories with positive tone (31 percent) than *El Mundo* (28 percent). Overall, both newspapers publish a large number of stories with both positive and negative tones, but as Figure 8.5 illustrates, there are important differences in the net tone used by both newspapers over time. This is explained because the tone is strongly related to the type of actor mentioned and the frame emphasized in the media stories. As expected,

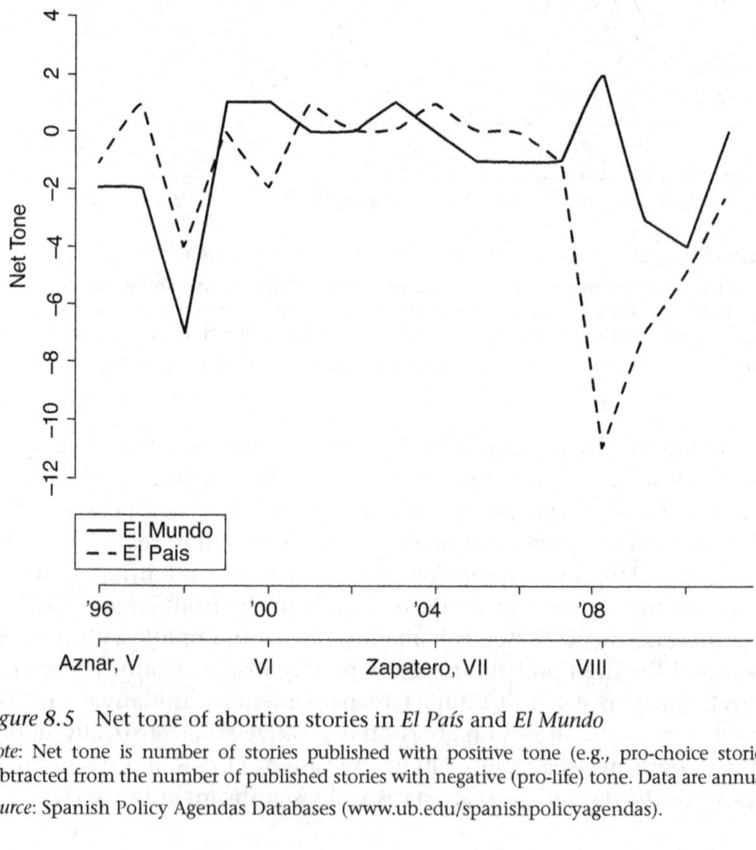

Figure 8.5 Net tone of abortion stories in *El País* and *El Mundo*

Note: Net tone is number of stories published with positive tone (e.g., pro-choice stories) subtracted from the number of published stories with negative (pro-life) tone. Data are annual.

Source: Spanish Policy Agendas Databases (www.ub.edu/spanishpolicyagendas).

El País and *El Mundo* frame abortion in different terms to promote a particular image of morality issues. *El País* more closely parallels the positions of the Far Left and the PSOE, and *El Mundo,* the Catholic Church and the PP.

As Figure 8.6 shows, the morality frame dominates in both media outlets, but it is especially important in *El Mundo,* being the main frame in 61 percent of the total stories on abortion. This illustrates the conservative profile of this newspaper like it does the high frequency of stories mentioning the Catholic Church as main actor compared to *El País* (Figure 8.7.b). In *El País,* the morality frame represents only

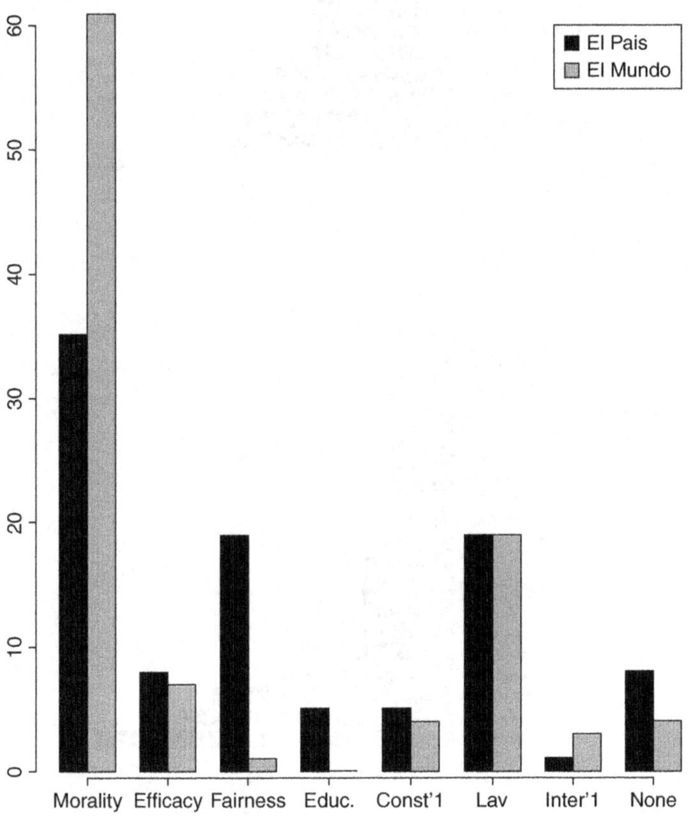

Figure 8.6 Main frame mentioned in abortion stories in *El País* and *El Mundo*

Note: For each newspaper the bars indicate the percentage of abortion stories using each of the frames indicated. Data cover the period from 1996 through 2011.

Source: Spanish Policy Agendas Databases (www.ub.edu/spanishpolicyagendas).

a. *El País*

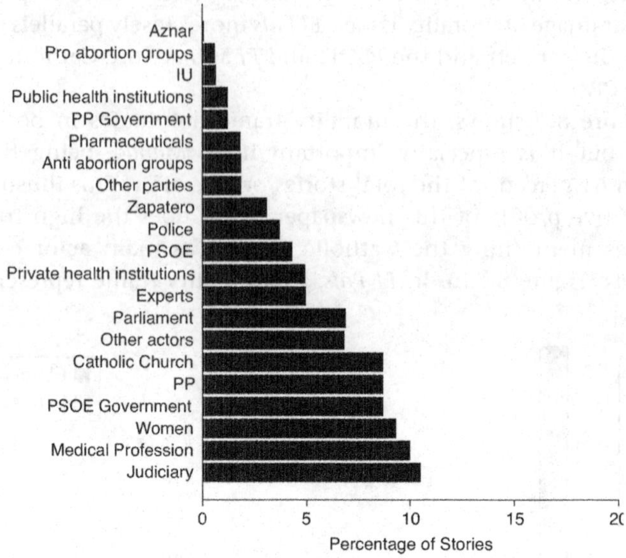

b. *El Mundo*

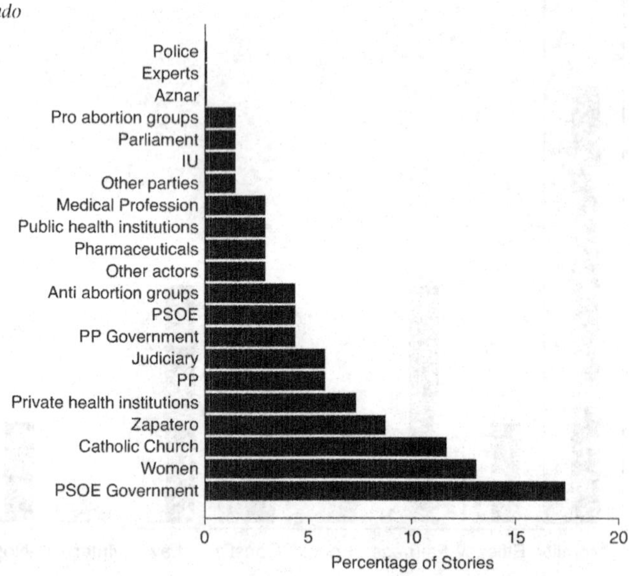

Figure 8.7 Main actors mentioned in abortion stories in *El País* and *El Mundo*

Note: Bars indicate the percentage of all abortion stories mentioning each of the actors or types of actors listed. Data cover the period from 1996 through 2011.

Source: Spanish Policy Agendas Databases (www.ub.edu/spanishpolicyagendas).

35 percent and the Church is given much less visibility. This newspaper uses a much more fragmented framing strategy, like Left parties do in the Parliament, emphasizing other aspects of the debate, especially those related to fairness. Fairness is the main frame in 16 percent of the total stories published in *El País* on abortion, while in *El Mundo* this frame represents only 1 percent (Figure 8.6).

The stories published by *El País* on the fairness frame are mainly associated with a negative tone and emphasize among other aspects the high percentage of abortions practiced in private health institutions (97 percent of the total), the access problems caused by the moral objection of doctors, and territorial inequalities among the regions of Spain. For example, in 1998, *El País* reported that in Galicia, Castilla y León, Extremadura, La Rioja, Navarra, Murcia, and Castilla-La Mancha there were no public health institutions practicing abortions. In *El País*, all the stories mentioning public health institutions as main actors have a negative tone. In contrast, in *El Mundo*, all are framed positively, emphasizing a completely different aspect of the problem, like the announcement that public health institutions would provide the morning-after pill for free, a good decision for those defending the pro-choice option but strongly criticized from conservative positions.

Overall, *El País* gives more visibility in its front pages to the *fairness* dimension compared to *El Mundo*, emphasizing inequality of access, especially through the public system, a framing strategy that strongly parallels the one followed by the PSOE and IU in the parliamentary arena and that was also supported by feminist movements and public health institutions, like the Instituto de la Mujer (Womens Institute) and interest-group organizations such as the Federación de Asociaciones en Defensa de la Salud Pública (FADSP, Federation of Associations in Defense of Public Health). This aspect of the problem goes unnoticed in the front pages of *El Mundo*, and it is nonexistent in the frame of the PP parliamentary initiatives.

The negative tone in *El País* stories is also associated with the *crime and law* frame, especially in 2008. To a large extent, this is explained by coverage of a highly visible court case against a doctor and his medical team that were accused of practicing illegal abortions; it came to light after a secret recording made by a TV program. *El País* gave much more coverage to this case than *El Mundo* and used it to denounce an "anti-abortion offensive" that ended with a private health institutions strike. These institutions decided to stop practicing abortions for a week in protest of the pressures received from anti-abortion groups denouncing, at the same time, the legal uncertainty suffered by doctors and women under the existing regulatory framework. *El Mundo* published only two stories about the case. The first emphasized that the accused had provided abortions to women in their seventh month of pregnancy. The second covered the sensationalist aspects of the story such as the cruelty with which the accused supposedly got rid of the aborted fetuses.

While the tone of media stories in *El País* was very negative in 2008 when this scandal occurred, *El Mundo* mainly published stories with a positive tone toward the pro-choice option (Figure 8.5), emphasizing the Socialists' plans of reforming the legislation passed in the 1980s toward more liberal positions. It openly referred to the "radicalization" of the government that was likely to reform the law even though most of the Spanish population (57 percent, according to this newspaper) supported the existing regulatory framework and the maintenance of the status quo, and only 36.5 percent were reported to support a "period model."

The different aspects highlighted by these newspapers within the *crime and law* frame are also illustrated in the actors mentioned in the stories. The judiciary and the medical profession are the actors most often mentioned in *El País* stories, while they have much less visibility in the front pages of *El Mundo* (Figures 8.7a and b). The police are also frequently reported as the main actor in *El País* stories, but never on the front pages of *El Mundo*. In both newspapers, most of the stories associated with the *crime and law* frame have a negative tone. However, in the case of *El País*, some are framed positively as they are used to give visibility to stories of exoneration and exculpatory sentences for doctors accused of practicing illegal abortions. Private health institutions are never framed with a positive tone in *El Mundo*, while they are in the case of *El País*, which uses them for giving visibility to good news, like the authorization of the first abortion clinic in Navarra that facilitates access to abortion to women that previously had to travel to other regions.

References to the *efficacy* frame also illustrate the different aspects of the problem emphasized by both newspapers. *El País* mainly uses it associated with a positive frame to give visibility to the authorization of the abortion pill. In *El Mundo*, it is used to criticize the inefficacy of a more liberal regulation, especially emphasizing fears related to an increase in the overall number of abortions that would result from the passage of a time-period model under the Socialist government. The constitutional dimension is used by both newspapers in a similar proportion of articles. In *El Mundo*, the constitutional frame is always associated with a negative tone highlighting the alleged unconstitutionality of the Socialist reform; *El País* uses it with a positive tone emphasizing the arguments given by some institutions, like the Council of State (Consejo de Estado) to support the constitutionality of the reform.

Overall, both newspapers emphasize different aspects of the abortion issue paralleling the strategies followed by Left and Right parties in the Parliament. Supporting their political allies is also clearly illustrated by the coverage of governmental actors and political parties in the front pages. In the case of *El Mundo*, the socialist government is the actor most mentioned in abortion stories (Figure 8.7.b). References to Socialist President José Luís Rodríguez Zapatero represent a large percentage of the total, compared to *El País*. Most of these stories are framed by *El Mundo* using a positive tone, reporting events that would be positively evaluated for those defending the pro-choice position but that would be unwelcome news for voters with a conservative profile. In contrast, almost all stories mentioning the PP or the conservative government in *El Mundo* are associated with a negative tone symbolizing the opposition of the Conservatives toward more liberal options. The same occurs in the case of *El País*, which despite having a smaller percentage of explicit references to parties and governmental actors, also supports the strategies of the Socialists and Left parties as the analysis of the framing of stories has illustrated.

Both *El Mundo* and *El País* report news in a manner consistent with the ideological beliefs and political orientation of their readers. As existing literature suggests, citizens, when consuming political information, avoid exposure to media outlets with an ideological position different than their own. They behave as "confirmation seekers," preferring to read positive news about the party for which they vote and negative news affecting the opposition party (Gentzkow and Shapiro, 2006; Newton and Brynin, 2001; Thompson, 2001; Pugglisi and Snyder, 2008). Given the conservative profile of *El Mundo* readers that are likely to vote for the PP, and the left orientation of those reading *El País* (see Chapter 5), both newspapers use the abortion debate to make a political point about their allied and opposing parties, but also a commercial one, responding to the preferences of their readers.

Conclusion

This chapter has explained the importance of framing in Spanish politics by analyzing the abortion debate in the parliamentary and media arenas. The exhaustive coding of every identified argument in parliamentary and media discussions of the abortion debate allowed us to show three things. First, parties on the Left and the Right emphasize different dimensions of abortion as a problem in order to promote or avoid regulatory reforms. Second, their framing strategies were supported by the two most-read newspapers in Spain, *El País* and *El Mundo*, following the strong political parallelism that characterizes the Spanish media system. These media outlets contributed to positive-feedback dynamics (Baumgartner and Jones, 1993), reinforcing tendencies already existing in other policymaking areas. Third, policy reforms cannot be attributed solely to framing strategies as opposed to other factors. Parliamentary groups give visibility to particular aspects of the debate to support their political preferences and political strategies. However, institutional factors, and more specifically control of government, are also important to explain policy dynamics in the regulation of abortion. However, Left parties did not enact progressive reforms immediately on coming into office for the first time, and Right parties have not immediately moved to eliminate these policies when they have come to office. So partisan control is mixed with a "sense of the times" as well. Abortion politics has evolved dramatically in Spain from the 1980s to present, and this has involved both a partisan logic as well as a larger cultural one.

Part IV
Conclusions

9
Transformations in Spanish Politics, 1982–2013

Throughout this book we have documented a number of important transformations in the functioning of Spain's democratic institutions since the transition to democracy. Spain has moved from a newly democratic state to a well-established democracy, characterized by a far-reaching welfare state. At the same time, a "normally functioning" democracy should not be read to mean a perfect one. Increased party polarization, a decreasing role of the legislature, a decline in mandate responsiveness, a great focus on a small number of policy issues such as the economic crisis, regional demands for increasing political autonomy or even secession, a continuous decline of media freedom, and a media system that encourages a government-opposition game in which highly publicized issues are considered largely for their value in the next election are important challenges for the functioning and health of Spanish democracy. So, while Spain has normalized as a democratic country, so has it developed many of the imperfections that are common in other Western democracies. We review these and other issues in this concluding chapter, beginning with the most important one, which is certainly not peculiar to Spain: the scarcity of political attention.

Attention scarcity

Political and media attention is scarcer than people realize. In an average month of 2004, the two leading newspapers devoted 50 percent of the stories on their front pages to just nine issues, including terrorism, political corruption, regional autonomy, the EU, political parties, the war in Iraq, immigration, government operations, and crime. Seven years later, in 2011, attention was equally concentrated on a few issues, with some variation in the topics but considerable overlap as well. Attention to war and immigration almost disappeared, as attention focused on the economic recession and related protests and social mobilizations against governmental responses to the economic crisis. The "usual suspects" – regional autonomy, crime, terrorism, political corruption, and governmental affairs – maintained their place among the stories that Spaniards read about virtually every month, year-in and year-out. While there may be no surprise that front-page attention focuses on just a few "major" issues, few have focused on the implications of this. One implication is that the vast majority of public policies see precious little attention. In this section we explore this characteristic of media coverage of politics, where Spain is certainly not much different from most other countries.

War, economic crisis, elections, terrorism, regional autonomy, and government affairs are certainly among the major priorities of Spanish citizens and political elites. However, they represent just a minor portion of the problems that become the object of significant political attention in the governmental and parliamentary arena; government officials are busy while the media is focused on topics 1 (macroeconomics), 2 (rights), and 3 (health) in making important decisions and implementing changes in public policies in dozens or hundreds of policy domains, far from the glare of media attention. Spanish newspapers and the media in general perform an important function as agenda setters, disseminating stories, ideas, and information about what is going on in the political sphere, and thus helping to reduce the asymmetries of information between political elites and the public. They also perform an important function as a watchdog of politics by expanding transparency in the public and private sectors, exposing corruption, maladministration, and corporate crime (Norris, 2001). The problem is that in doing so, the media gives unbalanced coverage of policy problems and politics. In our view, this is inevitable to some degree, but the degree of imbalance may be lower or higher. Some issues, by contemporary standards, are apparently not very newsworthy. But these topics may involve quite substantial investments of public funds and important policy decisions. Our point is simply that there is an economy of attention and by paying substantial attention to a few major areas, media systems in Spain (as elsewhere) inevitably leave most issues completely in the dark.

Scarcity of media attention is explained by journalistic norms, business logics, and the ideological biases of newspapers. Journalists write about what political elites are doing or planning to do and are affected by focusing events as well as changing social, economic, and political conditions. In doing so, they select those issues that are more rewarding in commercial terms and that correspond with the preferences of their readers; that is, to some extent they must follow a business logic designed to maintain or maximize audience share (of course there are journalistic norms that go beyond this as well). Whether they do it because of journalistic norms (which often emphasize conflict) or from a business logic, newspapers tend to emphasize bad news, highlighting the implications of policy failures, and focusing on the faults of their ideological enemies more than the strengths or accomplishments of those they support.

The level of coincidence of what the two major newspapers consider to be the most important issues of the day is striking. There are some different tendencies between the two newspapers: *El Mundo* pays more attention to political corruption and terrorism than *El País*, for example, whereas *El País* pays more attention to international affairs and has a more diverse agenda overall. But when one paper increases its attention to crime in a given week, so does the other; weekly correlations in levels of attention by policy topic across the two newspapers are extremely high. This means that there is agreement about when an issue has become more newsworthy. Largely, this is because both papers follow an indexing approach to the news (see Bennett, 1990); they report on what the government is doing, or talking about. Both newspapers pay attention to similar issues, highlighting what is wrong in politics, and paying special attention to those issues that generate the most confrontation among political elites and/or social actors.

The logic of what makes news implies, as an unstudied corollary to the obvious, that the vast set of issues on which government agencies are busily preparing new policies hardly get any attention in the front pages, and when they do get attention that attention may be focused on a particular set of highly contested debates. One of the most important things governments do, for example, is operate the public schools, and reforms, experiments, successes, and failures are constantly taking place in the educational domain. Education policy in Spain is, in fact, the single issue from within the social-welfare policy domain that gets the most attention in the front pages of the nation's newspapers. But this appears to be because education is linked with two of the most important cleavages of Spanish politics: religion and state-regional relations. Education coverage on the front pages of the newspapers, as in the case of other public social services like health, focuses on issues such as cuts in public expenditures and regulatory reforms, which generally involve mass demonstrations all over the country. However, in the case of education, attention is also focused on whether religion should be a compulsory course for undergraduates, whether the cost of teaching religion should be included in public education expenditures, and regarding the political autonomy of the regions to define their educational system. There are, of course, differences across the two newspapers in how these issues are presented, but both share the fact that their coverage of education focuses on just a few "newsworthy" elements of a very complex and important public policy. Almost 25 percent of the stories about general education in *El Mundo* are about state-regional relations, while the role of the Catholic Church and religion is only in 1 percent of the stories. In contrast, more than 10 percent and 12 percent of the stories in *El País* deal with religion and state-regional relations, respectively. The fact that Spain is one of the countries with the lowest education standards among advanced democracies occupies a marginal position in both front pages.

Thus, the logic of what makes something newsworthy is when that policy topic can be connected to one of the fundamental cleavages in Spanish electoral politics. Much of education policy has nothing to do with state-regional relations, and/or the role of the Catholic Church. But readers of the leading newspapers would be left with an impression that these are the fundamental issues facing teachers and school administrators across the country, and this is highly misleading.

Focusing events are also a key factor explaining scarcity of media attention. Some policy issues, from social-welfare service provisions to civil rights issues to the environment tend to be covered only when they can be linked to a major focusing event or to partisan arguments in the context of an election campaign. Agriculture only gets into the front pages in relation to food-safety scandals and the consequences of droughts, violence against women gets media attention especially after a new killing, and environmental issues are especially newsworthy when a catastrophe occurs or when they are linked to some human drama. Of course, the Spanish government, the EU, and regional governments are constantly engaged in important policy decisions in all of these domains, but they are surprisingly lightly covered, even in the "mainstream" or "serious" press.

Scarcity of media attention is also related to media coverage of particular political actors. Most of the time the media tends to focus its attention on what the government is doing or planning to do, giving special place to "official" sources, discounting opinions that are marginal or not represented in the halls of government, and giving priority to the routine reporting of the actions of government officials (Baumgartner and Chaqués-Bonafont, 2015: 6). The media selects among the issues political elites are talking about, and in doing so, they tend to stress the faults of their ideological enemies more than the virtues of their allies.

Newspapers follow completely different framing strategies in order to discredit and give a negative image of the rival party. We have provided several examples of this ideological bias. The analysis of political scandals illustrates that both newspapers pay more attention to the scandals of their opponents rather than their allies, as we discussed in previous chapters. *El País* gives more prominence to scandals affecting the PP and *El Mundo* to those of the PSOE. In 2009, *El País* devoted most of its attention to cover the Gürtel scandal, which directly involved the PP, paying almost no attention to PSOE cases of corruption like the Faisán case, which directly involved the socialist regional government in Andalucía (see Appendix B for a description of various political scandals in Spain). In contrast, in the same year, *El Mundo* fragmented its attention to political corruption, paying almost equal attention to a major scandal like Gürtel, than other scandals in which the PSOE was involved, like Faisán (Palau and Davesa 2013).

By the same token, *El Mundo* emphasized certain dimensions of abortion following the framing strategies of the PP, highlighting the morality frame and giving visibility to the Catholic Church and pro-life associations in line with the strategy followed by the conservatives in the parliamentary arena. If the socialist government is mentioned in *El Mundo* stories, it is to give a negative image of it in front of conservative voters, emphasizing aspects of abortion that would be negatively evaluated by PP voters, like initiatives to promote further liberalization in regulation. By contrast, *El País* uses a much more fragmented framing strategy, emphasizing especially the fairness dimension and mirroring the strategies followed by left parties and supported by feminist movements. The same occurs regarding religion and education, as we argued above. Both newspapers frame stories to attack their political opponents but also in a way consistent with the ideological profile of their readers following market-oriented strategies.

Globalization, internationalization of media groups, and increasing audience competition have not transformed the way newspapers report about politics; these effects neither increased the importance of the soft news in the front pages, nor created the high level of concentration of attention on just a few issues. Today, as two decades ago, most of the stories in the front pages of major newspapers have political content and provide relevant information about what is going on in politics. However, most policy decisions are not covered, discussed, or monitored by the media. One important implication is that the Prime Minister and the executive have great leeway to govern when the media are looking elsewhere. The greatest power of governments everywhere is to act with all the leeway and freedom that comes from a lack of oversight. This is especially problematic in a context of deep economic crisis and a decline of media freedom. The problem now is not only that the media depends on more and more of the financial support of large banks, multinationals, and governmental actors. In fact, during a period of economic crisis such as the current one, attention is so highly focused on the crisis and its response that other issues are virtually free from coverage, though policy activities continue in those areas, of course.

Journalists are not the only ones who must deal with attention scarcity. Political elites, in and out the governmental arena, also have to prioritize among issues, and this is partly explained by the fact that formal and informal institutions oblige them to do so. Formal rules impose important constraints on how political elites prioritize among issues when performing their tasks, setting substantial limits about who, how, and when to register an oral question or to introduce a bill. These formal rules directly define agenda capacity and thus affect how parties define their issue priorities. Overall, our results indicate the higher the agenda capacity (that is, for example, the number of oral questions per party, number of executive bills introduced in a single year, or quasi-sentences in the Prime Minister Speech), the greater the range of attention. That is, scarcity of attention is inversely related to agenda capacity. Whereas the "news hole" on the front pages of a daily paper is approximately seven articles per day, and this does not vary significantly over time or from paper to paper, the agenda capacities of different institutions of Spanish government differ dramatically.

In Chapter 4 we focused extensively on oral questions in Parliament, and here the agenda capacities differ from party to party. The fact that small political parties have ten times less capacity to register an oral question than statewide parties, explains why agenda diversity is larger for the PP and PSOE than any other party with political representation in the Spanish Parliament, as shown in Figure 4.4. When asking questions or introducing bills, parties like IU, CiU, or PNV tend to concentrate on a narrow set of issues, following a stable pattern across time, quite isolated from external events and changes in public opinion. This is especially the case for CiU and PNV, whose MPs keep asking questions and introducing bills dealing with regionalism. To these parties, with very limited choices on what to focus on, they focus on regional autonomy issues no matter whether an increasing number of Spaniards consider nationalism as the most important problem, with no matter what public perceptions may be about economic and political conditions, and no matter if they are giving support to the governing party or are in opposition. The same occurs in the case of IU, whose attention is highly concentrated on a small set of topics, especially after 2008, when the number of deputies declines from 8 to 2, which directly reduces its formal allotment of questions per session. In this sense, agenda capacity is the main factor that leads small regional parties to focus their attention mainly on regionalism; if they were larger and had more questioning opportunities, their attention would certainly broaden away from only their core issue of interest.

In sharp contrast to the small parties with the chance to ask only one or two questions per week, the statewide parties have the opportunity to ask many more questions. In this case, their agendas are of course much more diverse. But attention is still scarce and choices must be made. Our results show that for them their position in the government-opposition game determines their focus of attention. Both the PP and PSOE tend to register oral questions about a broader set of issues when they are governing as a means to give visibility to all type of governmental decisions, especially those that are more rewarding in electoral terms. Remember that the governing and main opposition party always have the same allotment of questions, and from 1996 to 2008, both pose about the same number of questions per session. While the governing party asks questions that cover the full range of government actions and try to bring attention to policy successes of the current government, the opposition parties tend to focus their attention on policy failures and highly conflictual issues such as regional autonomy, moral issues, and terrorism. There are some interesting variations between the PSOE (more confrontational on moral issues) and the PP (always more belligerent on terrorism). But for the most part, the logic is that either party, while in opposition, will use its question time to draw attention to issue domains that it avoided while in government.

In contrast to other countries, party conflict in and out of the parliamentary arena is not organized around the degree of state intervention in the economy and welfare state. Rather, in Spain, other issues related with long-standing cleavages in Spanish politics, mainly religion-secular values and center-regional relations, drive political conflict (Fernández-Albertos and Manzano, 2012). While in the opposition, both the PP and PSOE use their question time to force the party in government to define its policy position about these highly conflictual issues, for which there is an intense polarization among the electorate, and thus, with important political cost in case of political failure. To take a clear position about political decentralization, following increasing support toward recentralization among PSOE voters may be especially harmful for the PSOE, mainly because an important part of its electoral support comes from Catalonia, as discussed in more detail in Chapter 6. That is, the PP can exploit a potential internal cleavage for supporters of the PSOE by forcing the government to devote attention and clarify its position on an issue that divides its own electorate. In the same vein, to take a clear position about moral issues like same-sex marriage or abortion can be harmful for the PP, mainly because the PP attracts voters with quite different visions about the role of the state on the regulation of moral issues. Thus, the government-opposition game is played out in parliamentary question time with regional parties using their small share of time to focus on regional autonomy and the statewide parties highlighting those issues that divide the electorate of the governing party.

Opposition parties, when performing their task in the parliamentary arena, tend to follow the bad news reported in the front pages of *El País* and *El Mundo*. In fact, our results indicate that correspondence between media attention and oral questions is significant for the main opposition party, but not the governing party. This is, both the media and the main opposition party reinforce each other in this process of issue selection, emphasizing those issues than generate more confrontation and leaving aside those that create consensus. Governing parties, when asking questions to the executive in parliamentary question time, focus their attention on a wide variety of issues, emphasizing governmental policy decisions, especially on those issues that may be more beneficial for the executive.

In 2006, the PSOE asked questions about almost everything but immigration, though by that time most Spaniards had already come to consider this as one of the most important problems facing the country. In contrast, in the same year, more than 10 percent of the PP oral questions focused on immigration, almost one per session. In doing so, the PP is responding to public concerns, but it is especially following media attention. Both the main opposition party and the media reinforce their strategy of attacking the enemy. One of the main consequences is that there is a vast set of issues, even policy reforms enacted into law, taking place without much attention in the plenary meetings of the Spanish Parliament. Actually, a statistical correspondence between the number of executive bills passed and oral questions posed can be found only for two issues: civil rights and government operations (regional autonomy and public administration reforms). For all other issues correspondence is not significant, or even negative in the case of social policy, housing, energy, and agriculture. That is, oral questions, and likewise media coverage of policy issues, is typically unrelated to the areas where the Spanish government is enacting legislation.

One important implication of the disconnect between policy action and media attention is that the Prime Minister and the executive have great leeway to govern when the media are looking elsewhere. During the time when media attention is focused, say, on a corruption scandal, government agencies continue their work in hundreds of different policy actions, and it is safe to assume that many of these will affect the lives of millions of citizens for years to come. It matters greatly, in other words, that attention be allocated to those topics which deserve attention. Journalists must be active, and elected officials and other policy actors must strive to bring attention to issues that are being decided so that there can be an informed electorate and so that citizens can understand the actions of their governments. However, in Spain as in other countries, the various actors who might be expected to want to see attention to important policy debates each have perverse incentives. Political officials must stand for reelection and therefore may not want to embarrass their allies in government. Or if they are in the opposition, they may want to focus solely on faults of the government because this can help them during the next election. Journalists may want to write on important public policy stories, but they find that corruption scandals, points at which the parties are in dispute, or issues which are particularly embarrassing to the government are what their editors expect to see. In sum, there is no reason to assume that the most important policy debates will be rendered newsworthy just because they are important. At best, we see an imperfect allocation of attention where there is perhaps only a loose connection between the substantive importance of a policy issue and the likelihood that it will appear in the newspapers.

Democratic responsiveness requires attention; the system cannot operate without a vibrant and independent media sector including thousands of journalists on the ground working beats throughout society and in government agencies and working with equally vibrant civil-society groups. Political leaders need to search beyond the speeches and headlines as well to keep the government active in attempting to solve problems, no matter if their electoral allies will be frustrated by paying attention to a failure they might prefer to ignore. Ignoring failures to advertise successes while in government is equally misleading as ignoring success stories to focus only on faults while in the opposition. Democratic systems need information but have no systematic means to generate it, and may use incentives to suppress it based on the effect it may be expected to have on the next election. One result of the imperfections in the connections between public policy and political attention is that citizens are presented with what they often correctly perceive to be self-serving bits of information where neutrality and objectivity are given a backseat to the government-opposition game, with the media an active participant. Spain is not alone in seeing these dynamics at play in its daily newspapers, and its citizens are similarly not the only ones to become sometimes alienated from politics and rendered apathetic rather than engaged participants in the democratic process.

Attention scarcity and economic crises

The scarcity of attention is directly linked with economic conditions. Economic recession generates a major transformation in the content of the governmental and parliamentary agenda in two distinct ways. First, across the board, whether one consider the speeches of the Prime Minister, the oral questions in parliamentary oversight sessions, or the content of governmental bills, the diversity of the agenda declines under bad economic conditions. No matter which indicators we use to look at the economic situation (e.g., the number of people unemployed or citizens' perceptions about the economic situation), bad economic performance has a strong negative effect on agenda diversity. These effects are strongest for the case of the Prime Minister's speeches and for oral questions, with weaker effects for legislative bills.

These results corroborate existing analysis about agenda capacity (Baumgartner and Jones, 2015; Jones and Baumgartner, 2005). The capacity of political elites to adapt their policy priorities to changing political and economic conditions is larger in the case of the symbolic agenda than the decisional agenda. The Prime Minister has almost no formal restrictions on the content of his annual speech, and thus has full leeway to prioritize issues according to focusing events, of which an economic crisis would be among the most important. This is exactly what happened during the second legislature of Rodríguez Zapatero. In his last speech, he devoted almost 40 percent of the time to talk about the economy, five times more than in his first speech of 2008. The same occurs in the case of oral questions. MPs have a lot of formal restrictions in terms of the number of questions, but they have no restrictions regarding the issues they talk about when asking questions (with the only limit of being respectful with their peers). Analyses such as those by Jones, Larsen-Price, and Wilkerson (2009) for the US case or Chaqués-Bonafont and Palau (2011b) for the case of Spain give support to this finding. In both cases, results demonstrate correspondence between the political agenda and changing public opinion across issues is inversely related with institutional friction.

In contrast, in the case of bills, fragmentation of issue attention follows a steadier pattern. The capacity to adapt the decisional agenda to changing economic conditions is smaller, especially when the party in government does not have the majority of seats. To change policy goals is complicated when the executive depends on the support of other parties. This is the case of Rodríguez Zapatero and González in the early 1990s. Besides, as Adler and Wilkerson (2012) emphasize, the decisional agenda contains issues that the party in government cannot simply avoid. The so-called compulsory issues include not only bills about salient issues and present events but also legislation about expiring laws or provisions that have to be addressed along the legislature. As a result, the impact of economic conditions on the bills agenda diversity is not significant. So we see that some of the most visible, but also the most symbolic, agendas in Spanish politics are highly affected by a small number of focusing events (such as the state of the economy, but also by Spanish involvement in war, or other highly salient crises), but the lawmaking activity of the Spanish government goes on in all policy domains, no matter what. In this regard, Spain is very much like other Western democracies. The machinery of government continues, no matter what the Prime Minister may be speaking about or what the front pages of the nation's newspapers are focused on at the moment.

Overall, a reduction of agenda diversity means some issues are expelled out of the agenda. The economic recession generates increasing attention to economic-related issues, mainly macroeconomics, labor regulations, banking and commerce, and foreign trade, and a radical decline in the attention to welfare, rights, and the environment. This agenda-scarcity effect related to economic crisis is relatively consistent over time and across whether we look at speeches, oral questions, or legislative bills. Governmental issues are the only issue positively correlated with economic recession, and this is especially the case in the mid-1990s. The explanation relies on two different factors: first the debate about economic recession is always linked to the debate about crises of governability and the need to reform governmental structures in order to promote good governance. In the early 1990s, speeches, oral questions, and bills were related to the implementation of the New Public Management as the new paradigm to solve the crisis of governability and the importance to control political corruption. At the end of the first decade of the 2000s, political corruption comes back to the main stage of politics, although the debate about the good governance is now more focused on "transparency" measures and the implementation of new information technologies in the provision of public services. In any case, when the economy goes wrong, attention scarcity increases, keeping out of the political debate a vast array of issues.

Partisanship, elections, and attention

Policy preferences, of course, have a strong impact on how the government chooses to set its agenda and prioritize its actions across the various issues that could demand action (Budge and Farlie, 1983; Petrocik, 1996; Petrocik, et al., 2003). The increasing importance of personal rights issues in the governmental agenda from the mid-2000s is explained by the arrival in power of Rodríguez Zapatero in 2004. In less than a decade, regulation on abortion and same-sex marriage became much more permissive, religion was eliminated as a compulsory course in schools, and gender equality became significantly more prominent in the political debate. Policy preferences also explain increasing attention to health and water policy during the government of Aznar, and the prominence of foreign affairs under the mandate of González. Shifts in power at the top echelons of government clearly can be shown to alter the content of the governmental agenda.

However, the evidence provided in this book illustrates that while each Prime Minister can be associated with greater or lesser attention to individual issues, these issues are not always the same across leaders of the same political party. Rather, leaders of the same party but governing during different historical periods adopt substantially different issue priorities; meaning that at the end of the day, while individual leaders differ in their issue priorities, there are no consistent party-level differences in which issues are on the agenda during the period when left and right parties are in government. Our data (see, for example, Table 3.4) show that there are not important differences in attention across topics depending on which party is governing. Attention to issues does not significantly vary depending on the party that is giving the speech, introducing an oral question, or drafting executive bills. The PSOE does not have issue ownership on social affairs, health or the environment, and the PP does not monopolize attention to issues like the economy or government affairs, and this finding is consistent across all the institutional venues we have investigated. In fact, our results in Chapter 3 show that differences in attention are more closely linked to external factors (mainly the economy) than the partisan affiliation of the Prime Ministers.

These findings are perhaps surprising to those who assume that partisanship and electoral control determine the direction of the ship of state. In any case, based on the analysis presented throughout this book, we cannot predict which parties emphasize which issues across policy venues simply taking into account party ideology, but we can predict variations of the attention by looking at economic conditions. The PSOE, when performing its tasks in and out the governmental agenda does not tend to highlight welfare and the environment more than PP, and PP does not pay more attention to the economy than the PSOE. This is also the case when considering the content of party manifestos. During electoral campaigns, there are no significant and consistent differences in the amount of attention parties pay to issues or in the type of issues parties talk about, though there are differences in how they talk about these issues. Interestingly, these differences in party position are not particularly intense regarding welfare issues, as occurs in other countries, but rather the parties in Spain differ especially starkly on issues regarding individual rights (Fernández-Albertos and Manzano, 2012). Partisan and ideological conflict over welfare is strikingly small. Parties can afford to have similar policy positions, supporting an increasing expansion of the welfare state in a context in which partisanship is a poor predictor of welfare-state preferences. The PP and PSOE electorates' preferences toward welfare are very similar, and this is mainly explained by the fact that the expansion of the welfare state in Spain is compatible with a nonredistributive agenda.

Parties focus their attention on some issues and not others taking into account economic and social conditions, the type of government, and their institutional status in the government-opposition game. Elections allow governments to enact ideologically driven policies, but political leaders have to adapt the political agenda to changing social, political, and economic conditions and also to the opportunities provided by the institutional context. The outbreak of an economic crisis, an environmental catastrophe, or a food-safety scandal may force policymakers to attend issues they would otherwise try to avoid. Aznar paid more attention to agriculture issues than González and Zapatero, not for reasons related to party preferences but because it had to deal with the mad-cow scandal, a food-illness outbreak initiated in the UK that arrived in Spain at the end of 2000 causing high media attention and concern on food safety.

Attention to immigration by the PP and PSOE may respond to different ideological positions regarding the impact of immigration flows on the economy and society but attention to this issue cannot be explained without considering the continuous growth in immigration coming from Latin American and African countries that took place in Spain, as the percentage of immigrants in Spain increased from 1.6 percent in 1998 to 12 percent in 2012.[1] However, there is no greater factor in politics than economic conditions to change the policy focus of a governing party. During times of economic growth, Prime Ministers may focus on what they prefer. But in times of economic crisis, they have no choice but to make this their overwhelming priority. In 2004, Rodríguez Zapatero devoted less than 5 percent of his speeches to talking about the economy. For the first time, welfare-related issues captured more attention in the first speech of the Prime Minister than economic issues. By 2011, the economy was the focus of more than 40 percent in the last speech of the PSOE leader, as he shifted substantially to this issue, leaving aside some of the milestones of his mandate and many of his previous areas of focus.

To implement policy preferences, political elites also take into account public mood and changing social and religious values. The conservative Rajoy government was forced to take a step back in its attempt to reform the abortion law toward a more restrictive position, mainly because today, most Spaniards give support to the legalization of abortion. On abortion, most citizens' positions are fixed, as it is an issue associated with strong religious beliefs. But over time, the increasing secularization of society has meant that citizens' preferences are increasingly moving toward positions that could not have been sustained in the 1980s. So policy proposals shift not only in response to economic crises and opportunities or focusing events, as with the economy, but also in response to long-run trends in public opinion and the demographic makeup of society.

Finally, policymakers' capacity to implement ideologically driven policies is also influenced by institutional factors, like variations in the type of government or the consolidation of a multilevel governance system. The minority governments of the 1990s, where regional parties (CiU but also the PNV and CC) were pivotal for the government formation avoided also the type of recentralizing policies that were later implemented by the conservative government of Mariano Rajoy. The reforms introduced on education or the refusal to negotiate a new financing system for Catalonia would not have been possible if the conservative had continued to depend on the support of regional parties for passing legislation in the Chamber. So, governmental capacity to enact ideologically driven policy reforms are often conditioned on having unilateral control, which in Spain is not to be taken for granted.

Overall, party preferences cannot predict variations in issue attention, and this is a common pattern across different venues. Party leaders differ in their policy preferences but their capacity to implement them depends to a large extent on external factors such as the state of the economy, long-run social trends such as demographics and public opinion, and institutional factors including minority government status and, over time, the consolidation of a multilevel system of governance. As we argued throughout the book, this has important implications on mandate responsiveness.

Long-term trends in Spanish politics: responsiveness

One of the clearest findings from our research is that policymakers in Spain are increasingly less capable to translate their policy promises into policy decisions. This is often the result of institutional factors (such as government with minority status) or unexpected circumstances like the outbreak of an economic or foreign policy crisis which demands attention even if the electoral campaign had focused on promised reforms in other areas. Another long-term process strongly detrimental to mandate responsiveness in Spain is the progressive consolidation of a multilevel governance system. Our results give support to existing theories about mandate responsiveness. In Spain, under minority, the capacity of the governing party to fulfill its mandate declines significantly, no matter which party is governing. Felipe González had to renounce some of his electoral promises, especially those related to individual rights, as a result of his agreement with regional conservative parties; Aznar had to introduce the economic fiscal reform as one of his priorities after winning elections in 1996 as a result of the agreement with CiU. In both cases, results significantly illustrate that the ability of party leaders to fulfill their electoral promises decreases when they must ally with others to pass legislation (Budge and Keman, 1990; Klingemann et al., 1994).

Correspondence between mandates and policy outcomes also depends on economic conditions. Bad economic conditions are an important part of the explanation for the decline of mandate responsiveness in the case of the PSOE governments of Felipe González and Rodríguez Zapatero. The outbreak of the current economic crisis, for example, has imposed a redefinition of agenda priorities and preferences against most of the electoral promises made by both the socialists and the conservatives. In the wake of the recession and collapse of the country's housing market, José Luis Rodríguez Zapatero's socialist government implemented severe austerity measures aimed at reducing state spending and limiting the public-sector deficit; none of these initiatives were defined in its party manifesto. Cutbacks were taken in social welfare and pension programs, and labor-market reforms were implemented because of the crisis, though these were certainly not among the electoral promises of the government before its arrival in power. Mariano Rajoy's conservative government continued to implement policies to curb spending and to control the state deficit, launching a major reform of the labor market and financial system, and raised taxes, contradicting also some of their ideological principles and electoral commitments concerning light regulation of the financial markets and low taxes.

Changing institutional structures and external events affect the capacity of the executives to fulfill their electoral promises. However, these factors cannot explain why mandate responsiveness is decreasing over time. We argue this is linked with the consolidation of a multilevel governance system (Chaqués-Bonafont, et al., 2013). From 1982 to 2011 there is a steady gradual increase in the number (and percentage) of issues that are mentioned in the party manifesto as important issues but never receive any attention either in the speeches or executive bills, as we showed in Chapter 3. There are an increasing number of issues highlighted during the electoral campaign that later on are left unattended by the governing party. Our results indicate that this lack of correspondence is increasing across time, and is especially important for those issues with shared jurisdiction between the national, regional, and EU governments.

On the one hand, the Spanish government has to implement EU-policy decisions that in some occasions may be completely unrelated to Spanish politics, and thus never considered in any party manifesto or even discussed in Spanish newspapers to any significant extent. Problems about seal hunting are never a hot issue in Spain, just as the regulation of linen or fishing is not an electoral issue in Luxembourg or in Austria. But all countries have to adapt their legislation to these issues, maybe even to pass executive bills about issues that reflect no previous ideological position-taking. Further, executives throughout the EU must transpose these directives regardless of whether it agrees with the content of the EU legislation, and regardless of whether the EU decisions will bring net costs or benefits to the country in question (see

Zubek, 2011). Most of the policy measures recommended by the EU to overcome the current economic crisis were implemented in spite of the high costs they imposed in terms of reelection to the incumbent. On the other hand, the Spanish government may be paying attention to issues whose jurisdiction is in the hands of other levels of government, and thus whose final decision depends on the agreement with other governments.

Our argument is that the capacity of governmental actors to fulfill policy promises for these issues may be lower for two reasons. First, institutional friction, that is, the cost to reach an agreement and promote policy change, would be larger for issues with shared jurisdiction; and second, the incentives of parties to start a strategy of blame-shifting and credit-taking is larger for these issues than the rest. The PP government of Aznar could not implement the PHN mainly because the EU opposed financing the construction of this macro infrastructure. By the same token, the government of Mariano Rajoy could not fulfill one of its electoral promises, the construction of a high-speed train that crossed the Pyrenees by the middle, the so-called *Corredor Central*, mainly because the EU is opposed to it. In fact, the PP, when campaigning in the last general election already knew the EU decision on the high-speed train was final, but it kept pushing forward this issue as one of its main electoral promises in order to maximize electoral rewards. It could later blame the restrictions of the EU for its inability to deliver on this promise, making an electoral focus on this issue virtually cost-free for the party, but also without benefit if it believed that fulfilling this electoral promise would be better for its supporters than promising something it knew could not be delivered.

Political parties may follow a strategy if making false promises, or ones that they cannot unilaterally keep, in a multilevel governance context, but this was more difficult before the rise in power of the regions or of the EU. Today, the costs of the lack of implementation of policy promises are especially low (Anderson, 2006; Soroka and Wlezien, 2010; Lord and Pollack, 2010; Bellamy and Castiglione, 2013). Spanish citizens have a rather vague idea of which level of government is politically accountable for each policy area (Lago and Lago, 2011; Gunther and Montero, 2009), which creates strong incentives for policymakers to keep paying attention to these issues in the party manifestos in order to respond to the wishes and concerns of the electorate, knowing that in case of inaction they can always blame the EU or the regions if they fail to deliver (Chaqués-Bonafont Palau and Muñoz, 2014).

In all, delegation of powers upwards toward the EU and also downwards toward the regions throws into question the problems of the representative model of democracy, and especially the need to reform democratic institutions in order to reinforce responsiveness (Schmidt, 2006). The capacity (or willingness) of policymakers to respond to political (electoral) promises is declining across time, and this is related to, at least to some extent, the consolidation of a new institutional framework in which the limits of political responsibility are blurring. This generates problems of political accountability and fosters political discontent and alienation. This is especially relevant in a context in which major decisions, even a reform of a constitution, are driven by EU decisions.

Fragmentation of power and the new politics of conflict

The Spanish executive, as in other majoritarian democracies, has a predominant role in Spanish politics as agenda setter and legislator that has been reinforced through the process of Europeanization. Membership to the EU has bolstered the already dominant position of the executive vis-à-vis the legislative. On the input side of EU policymaking, the Spanish Parliament and regional governments play a marginal role in EU decisions. As explained in Chapters 6 and 7, the Spanish Parliament and the regions have little capacity to get information or monitor the government regarding EU decisions. By the same token, the Spanish Parliament only intervenes in the transposition of a minor part of EU regulations and policies into the domestic setting. This is illustrated by the fact that in Spain, most directives are transposed by secondary legislation (mainly decrees), neglecting the role of the Spanish Parliament, and this has been consistent since 1986 (Palau and Chaqués-Bonafont, 2012).

The delegation of competences upwards toward EU institutions has had a great impact on the relative powers of executive and legislative branches, but the increasing neglect of the Spanish Parliament in the policymaking process results also from domestic factors, and more specifically from variations in the policy style of government. As we have argued in Chapter 3, the politics of consensus that characterized Spanish democracy during more than two decades has shifted toward increasing polarization and confrontation. One of the factors leading this change in the policy style of governing is the increasing use of initiatives and policy instruments that neglect the role of the Parliament in the legislative process. Since the transition to democracy the executive has always dominated the policymaking process, being responsible for the initiation of most of the legislation passed in the Chamber, but during the last decade, the increasing use of decree-laws to take decisions, even on issues that have nothing to do with urgent necessities, and the use of non-ordinary procedures to pass legislation, have further reinforced the power of the executive over the legislature in Spanish politics.

Opposition groups have reacted to this increasing neglect of Spanish Parliament with increasing confrontation and opposition to governmental initiatives. The mean percentage of negative votes for legislation has reached unprecedented levels, especially since the outbreak of the current economic crisis and the arrival to power of the conservatives, governing with absolute majorities of seats since 2011 (Palau, Muñoz and Chaqués-Bonafont, 2014). The increasing polarization of Spanish politics, however, is a long-standing trend and already showed up during previous legislatures, when Aznar was governing with absolute majority, from 2000 to 2003, on issues related to deep and long-lasting political cleavages like religion. In the early 2000s, regulatory change on moral issues and education altered the predominant position of the Catholic Church in Spanish politics. To liberalize abortion, to legalize same-sex marriage, or to permit the therapeutic use of "savior siblings" are all policies strongly opposed to the precepts and values of the Catholic Church, but as we saw in Chapter 8, these have come to pass. By the same token, to eliminate religion as a compulsory course and to question the role of religion in education means to alter one of the most controversial but also long-lasting agreements made during the period of democratic transition.

But polarization and political confrontation are especially acute for the case of political decentralization. The percentage of citizens in favor of changing the present model of territorial distribution of power in Spain has increased exponentially from the late 2000s. Opinion has not shifted only in favor of secession, but at the same time also in favor of recentralization; that is, opinion has dramatically polarized with more supporting dramatic change on one side or the other, and fewer supporting the status quo. Large demonstrations for an independence vote in Catalonia are accompanied by increasing visibility of the issue in the media outlets and that frame stories relating to the regions using an increasing negative tone. As the debate has become increasingly intense over time, political decentralization has become one of the preferred weapons of political parties to attack each other in search of electoral rewards. While neither seems able to make a clear proposal about how to solve the issue, each can attack the other while they are in opposition that the governing party has negotiated poorly on behalf of the central state, and thereby use the issue of regional autonomy as a means to maximize electoral rewards.

Thus, Spanish politics has increasingly become polarized on the regional issue even while it has not engaged with any significant debate on the role of the EU. Opinion polls show that following the outbreak of the economic crisis citizens are increasingly critical toward EU institutions and less likely to believe that Spain benefits from European integration as much as before. The crisis has sapped support for EU institutions, especially the European Central Bank, and seems to have significantly affected the output-based form of legitimacy that has traditionally upheld support for European integration in Spain (Jones, 2009; Díez Medrano, 2007). Still, as of today, we have seen no serious discussion of leaving the EU and political debate over European integration is almost inexistent.

Interestingly, this major transformation of power within the Spanish national government has occurred without any significant growth in partisan conflict on EU issues or the development of any strongly euroskeptical parties. Increasing pooling of sovereignty in European integration has not resulted in increasing awareness, mobilization, and polarization over the EU as it has occurred in other countries (De Wilde, 2011; Hooghe and Marks, 2009). As we argued in Chapter 7, the economic crisis has increased attention to EU-related economic decisions from opposition parliamentary groups, but these questions were not oriented to openly criticize EU policies but rather to use the EU frame to attack the incumbent and obtain electoral rewards.

In short, in this book we have explained the transformation of Spanish politics from a model of consensus politics to a new model characterized by increasing confrontation, and the gradual decline of mandate responsiveness. Furthermore, political parties look for neither the accommodation of different views and perspectives about what should be done, nor an open debate in the media and/or parliamentarian arena where all political forces participate to reach common policy positions. By the same token, we demonstrated that the Spanish executive rarely adheres to its electoral promises, and this is especially true in cases for those issues with shared jurisdiction with the EU and/or the regions. These changes in the quality of democracy have not gone unnoticed by Spanish citizens. Political trust in political institutions has declined exponentially, and for the first time from transition to democracy, opinion polls seem to indicate the end of bipartism, and the growing importance of new forms of political mobilization. Our goal in the future is to analyze these transformations in the political, social and economic context and the consequences they may have in agenda dynamics over time.

Glossary

Board of Spokespersons	Junta de Portavoces (Spokespersons of parliamentary groups make up the Board of Spokespersons)
Congress of Deputies	Congreso de los Diputados
Investiture speeches	Discurso de Investidura
Ombudsman	Defensor del Pueblo
Parliamentary Bureau	Mesa del Congreso (Bureau responsible for the management of the Chamber vested with the collegial representation of it).
Prestige	Environmental catastrophe caused by the sinking of the Prestige oil tanker in 2002
Regional statutes	Estatutos de Autonomía
Senate	Senado
Spanish Courts	Cortes Generales (Bicameral Parliamentary Chamber composed of the Congress of Deputies and the Senate)
State of the Nation Debate	Debate sobre Política General en torno al Estado de la Nación

Appendices

Appendix A: Subtopics of the Comparative Policy Agendas Topics

100. General economics
101. Inflation and interest rates
103. Unemployment
104. Monetary policy, the Bank of Spain (National Central), monetary reserve discount rate
105. Budgets and spending budget law
107. Taxes, tax policy, and tax reform
108. Industrial policy
110. Control and stabilization of prices
199. Other economics
200. General rights
201. Ethnic minorities and racial discrimination
202. Gender discrimination and rights; homosexual discrimination and same-sex marriage.
204. Age discrimination
205. Discrimination of people with illnesses or disabilities
206. Voting rights, political participation, and representation
207. Freedom of expression, assembly, and demonstration; pornography; equal rights in general
208. Right to life
209. Activities against the state
212. Religious freedom in general, Catholic Church, and other religions
225. Human rights
230. Immigration and refugees
299. Other rights
300. General health
301. General reforms of the National Health System (NHS)
302. General questions about the coverage of the NHS
321. Regulation of the pharmaceutical industry
322. Sanitary facilities; hospitals
323. Agreements between the NHS and private companies
324. Medical malpractice, malpractice, and compensation systems
325. Human resources, education, and training
331. Disease prevention and health promotion
332. Child health
333. Mental diseases

Continued

334. Long-term treatment, rehabilitation services, hospice, and problems relating to aging
335. Pharmaceutical expenditure, government consumption, and drug prices
336. Coverage of fertility treatments, dentists, etc.; accessories like glasses, wheelchairs, etc.
341. Tobacco
342. Alcohol
343. Control of illegal drugs
344. Generic issues related to illegal drugs, tobacco, and alcohol
398. Research and development in health
399. Other health
400. General agriculture
401. Agricultural trade
402. Subsidies and agricultural regulation
403. Food inspection and safety
404. Agricultural marketing and promotion
405. Animal and crop disease; pest control
406. Sanitary and living conditions of farm animals
407. Environmental issues related to agriculture
408. Fishing and hunting policy
498. Research on agriculture and livestock
499. Other agriculture
500. General labor
501. Working conditions, work accidents, and compensation schemes
502. Employment training and workforce development
503. Pensions and early retirement; other employee benefits
504. Labor unions
505. Employment policy and collective negotiation
506. Youth and employment
507. Unemployment and disease compensation
510. Social Security; general issues
511. Employment and unemployment in specific industries
520. Fraud and abuse of the Social Security system
529. Labor and immigration
599. Other labor
600. General education
601. Higher education
602. Elementary and secondary education
603. Education underprivileged students
604. Professional training
606. Special education for disabled students
607. Libraries and education excellence
609. Cultural policy
610. Sports
612. Cultural patrimony
698. Research on education
699. Other education
700. General environment

Continued

701. Water quality
703. Waste disposal
704. Hazardous and toxic waste
705. Air pollution, global warming, and noise pollution
707. Recycling
708. Indoor environmental hazards
709. Species and forest protection
710. Pollution and conservation of the coast
711. Land conservation
712. Territory planning
798. Research and development
799. Other environment
800. General energy
801. Nuclear energy
802. Electricity and hydroelectricity
803. Natural gas and oil (including offshore oil and gas)
805. Coal and mining
806. Alternative and renewable energy
807. Energy conservation
898. Research and development
899. Other energy
1000. General transportation
1001. Mass transportation and safety
1002. Highway and roads construction, maintenance, and safety
1003. Airports, air traffic control, and safety
1005. Railroad transportation and safety
1006. Automobile industry: traffic and regulation
1007. Maritime issues and naval industry
1010. Public works (infrastructure development)
1098. Research and Development in transportation
1099. Other Transportation
1200. General Crime
1201. Police and crime-fighting authorities
1202. Organized crime and financial-related crime
1203. Illegal drug production, trafficking, and control
1204. Judiciary system
1205. Prisons
1206. Youth crime and antisocial behavior
1207. Child abuse and child pornography
1208. Domestic violence
1209. Weapons control, private security forces, and civil protection
1210. Criminal and civil code
1211. Crime prevention
1212. Tax fraud
1230. Civil disorders, riots, and civil wars
1260. Terrorism and counterterrorism
1297. Ordinary crime
1299. Other crime

Continued

1300. General social welfare
1301. Nutrition programs and food assistance
1302. Poverty and assistance for low-income families
1303. Elderly issues and elderly assistance programs
1304. Assistance to the disabled and handicapped
1305. Social services and volunteer associations
1308. Work-life balance and child care
1399. Other social welfare
1400. General housing
1401. Urban housing
1403. Urban economic development and general urban issues
1404. Housing policy in rural areas
1405. Rural economic development and general urban issues
1406. Public housing programs
1408. Elderly and handicapped housing
1409. Housing assistance for homeless and homeless issues
1411. Housing market and real estate speculation
1499. Other housing
1500. General commerce
1501. Banking system and financial institution regulation
1502. Securities and commodities regulation
1504. Mortgages, credit cards and other services of the credit market
1505. Insurances
1507. Bankruptcy
1520. Antitrust regulation
1521. Small and medium enterprises issues
1522. Copyrights and patents
1523. Domestic disaster relief (natural disasters and accidents)
1524. Tourism
1525. Consumer safety, consumer fraud, and data protection
1526. Lottery and gambling
1529 Professional and business associations
1599. Other commerce
1600. General defense
1602. NATO and other defense alliances
1603. Military intelligence, CIA, espionage
1604. Military readiness and capabilities
1605. Arms control and nuclear nonproliferation.
1606. Military aid and weapons sales to other countries
1608. Human resources and military personnel
1610. Military procurement and weapons system acquisitions
1611.Military installations, properties and building construction
1614. Military environmental compliance
1615. Civil protection services and Armed Forces
1616. Consequences of military activities for civilians
1617. Defense contracts
1619. Direct participation in war
1620. Human rights' violations in war; relief of claims against military forces

Continued

1698. Research and development on military issues
1699. Other defense
1700. General science, space, and technology
1701. Space missions and research
1704. Satellites and other space technology with commercial use
1705. Science technology transfer and scientific international cooperation
1706. Telecommunication and telephone services
1707. Media
1708. Weather forecasting and geologic issues
1709. Computer industry and computer security
1798. Research and development (projects and resources)
1799. Other science, space, and technology
1800. General foreign trade
1802.Trade agreements, disputes, and regulation
1803. Export promotion and regulation
1804. Overseas private investment and Spanish investments abroad
1806. Productivity and competitiveness; Spain balance of payments
1807. Imports and regulation of imports
1808. Exchange rates
1899. Other foreign trade
1900. General international affairs
1901. Foreign aid
1902. International agreements on environmental issues
1905. Developing countries issues
1906. International finance system and economic development organizations
1907. China, India, and Pakistan
1908. Russia and Former Republics of the Soviet Union
1909. Eastern Europe
1910. European Union (all institutional issues)
1911. Sub-Saharan Africa
1912. South Africa
1913. Western Europe
1914. Central and South America
1916. USA
1917. Canada
1919. Japan, Koreas, and Oceania
1920. Middle East
1921. Southeast Asia
1922. Maghreb
1925. Human rights
1926. International organizations
1927. International terrorism
1928. International nongovernmental organizations
1929. Diplomacy
1930. Mediterranean
1999. Other international affairs
2000. General government operations
2001. Intergovernmental relations and local government

Continued

2002. Public administration efficacy
2003. Postal service
2004. Civil service
2005. Appointments and salaries
2006. Awards and public honors
2007. Government procurement, contracts, and corruption
2008. Privatization of public sector and nationalizations
2009. Central Tax Administration
2011. Parliament and Constitution
2012. Regulation of political activities, elections, and election campaigns
2015. Claims against the government
2030. National holidays
2040. Transition and consolidation of democracy
2050. Political independence, decolonization, and state building
2060. Political parties as organizations
2070. Government-CCAA relations
2099. Other government operations
2100. General public lands
2101. Natural parks and protected areas
2103. Use of natural resources
2104. Water and sea resources: development, public works, and harbors
2199. Other public lands
2300. General culture
2301. Cinema, theatre, music, and dance
2302. Publication of books and literary works in general
2399. Other culture
2700. Weather
2900. Sports
3001. Natural death notices
3002. Violent death notices
3099. Other death notices

Note: Topics 2300 and above are used only in media, as there are few to no governmental activities on these topics.

Appendix B: Brief Description of Main Corruption Cases in Spain

GAL: Scandal about the involvement of the socialist government in the creation and coordination of the GAL (Anti-terrorist Liberation Groups (Grupos Antiterroristas de Liberación), who were responsible for the so-called "dirty war" against terrorism.

CESID: A number of scandals linked to the CESID (Superior Centre for Defense Information (*Centro Superior de Información de la Defensa*), which took place from 1977 until 2002. This included the "CESID papers" scheme and the cases about illegal bugging of other parties and those in senior positions.

FONDOS RESERVADOS: Scheme involving the illegal use by the socialist government, between 1987 and 1993, of reserved funds of the home office for fighting terrorism and drug trafficking.

FILESA CASE: Network of companies accused of illegally financing the PSOE.

JUNTA DE ANDALUCÍA: A number of schemes based on accusations of corruption and influence bartering against the Andalusian Regional Government (*Junta de Andalucía*), presided over by the socialist Manuel Chávez between 1990 and 2009.

BERMEJO: Case about the misappropriation of funds by former Justice Minister and socialist Mariano Fernández Bermejo. It also includes the scandal that appeared following a joint hunt with Judge Baltasar Garzón in 2009, in which he was accused of violating the principle of separation of powers.

CNI: A number of accusations directed at socialist politician Alberto Saiz, director of the National Intelligence Centre (*Centro Nacional de Inteligencia* (CNI)) between 2004 and 2009.

SOGECABLE: Dispute between *El País* and *El Mundo* in their bids for media channels Canal Satélite Digital and Vía Digital, respectively, and the support that each of the platforms received from the PP and PSOE.

FAISÁN: Judicial investigation initiated by Judge Baltasar Garzón that uncovered an extortion plot by ETA based in the Bar Faisán in Irún. It became political in nature when the PSOE government was accused of allowing the May 4, 2006, tip-off that warned the owner of the establishment that a police raid would be taking place.

GESCARTERA: Financial scandal that came to light in 2001 when a company called Gescartera was audited by the Spanish Securities and Exchange Commission (CNMV). The case became political in nature when Rodrigo Rato, a PP politician, was accused of being implicated.

GÜRTEL: Political corruption case linked to the PP that came to light in February 2009, led by businessman Francisco Correa, whose companies were accused of fraudulent use of public funds. It included the Valencian plot, which involved politicians and senior members of the PP.

BALEARES: Includes the "Sóller Tunnel case" in which PP politician Gabriel Cañellas, president of the Regional Balearic Islands Government between 1979 and 1995, was accused of bribery and corruption. This also included corruption cases linked to PP politician Jaume Matas, president of the Regional Balearic Islands Government between 1996 and 1999, and between 2003 and 2007.

TRABAJO: Corruption scandal related to the performance of Employment and Social Affairs Minister (*Ministro de Trabajo y Asuntos Sociales*) and PP member Eduardo Zaplana, who was in office between 2002 and 2004. He was accused of embezzlement of public funds, giving preferential treatment to individuals and companies, and misappropriation.

ÁLVAREZ-CASCOS: Politician Francisco Álvarez- Cascos was accused of corruption in his performance as Minister of Transport and Infrastructures of the PP government between 2000 and 2004.

PIMENTEL: Scandal caused by the abandonment of politics in 2000 of former Minister of Labor and Social Affairs of the PP between 1999 and 2000 as a result of the alleged irregularities committed by Juan Aycart.

ZAMORA: Corruption scandal linked to the PP that came to light in 1997, following accusations of the illegal collection of fees by former leaders of Zamora-based PP and constructor Eduardo Corrales.

TOMEY: PP senator and president of the Guadalajara Provincial Council Francisco Tomey was accused in 1998 of allegedly falsifying documents.

LINO: Scandal that came to light in 1999 about alleged irregularities in the payment of EU subsidies for the cultivation of flax during the PP's time in office.

ERCROS: Scandal linked to irregularities committed in the sale of Ertoil, which was carried out when former minister, PP MP Josep Piqué, was advisor to the Ercros group.

COMUNIDAD DE MADRID: Series of scandals that affected the PP local and regional government of Madrid from the beginning of 2000, including the case of the Madrid mortuary and the case of the turncoat deputies, Tamayo and Sáez.

MARBELLA: Corruption scandals that were linked to the government of Jesús Gil, mayor of Marbella from1991 until 2002, under the banner of his namesake party, GIL; and also the corruption cases that affected other members of the town hall, such as the "Malaya Case."

PRETORIA: Corruption scandal uncovered in October 2009 for supposed offenses involving bribes, corruption related to town planning issues, money laundering linked to politicians of the PSC (Catalan socialist party) and CiU (Catalan party Convergence and Union).

HACIENDA: Corruption case involving senior officials of the Inland Revenue and businessmen accused of falsifying documents and bribery.

ESTEVILL: The Higher Court of Justice of Catalonia (*Tribunal Superior de Justicia de Cataluña*) in 1996 initiated investigation proceedings against former Judge Pascual Estevill for supposed crimes of bribery.

Appendix C: Political Decentralization

In order to capture whether each of the political activities measured in the Spanish Agendas Project is related to the process of political decentralization (or regionalization), we have created a categorical variable (no regionalization = 0; regionalization = 1). That is, we have analyzed for each bill, oral question, or speech whether it is related to the capacity of regions to exercise authority in their territory (self-rule), or to modify the participation of regions in the policy-making process at the national level (shared rule). In the case of speeches, this includes all the explicit references (measured as quasi-sentences) in the speech the Prime Minister to the *Estado de las Autonomías*; in the case of bills, the variable is coded 1 for all bills that explicitly relate to one or more regions introduced by the government (for governmental bills) or introduced by political parties or the regions (in the case of parliamentary bills). Bills designed to establish the principles and basic regulation of issues with shared jurisdiction (framework laws, or *leyes de bases*) like health, or education, are coded 0. In the case of oral questions, we have coded 1 for all the oral questions introduced in plenary meetings that explicitly relate to political decentralization and specific problems of a region. Finally, the data about party manifestos relies on the Comparative Manifestos Project (www.cmp.org), and similarly codes 1 for any platform relating to regions.

From here, we have classified all political activities related to political decentralization into five categories and 21 subcategories (see Table 6. A1). This classification departs from the regional authority index developed by Hooghe, Marks, and Schakel (2008). The first two categories, policy scope and fiscal autonomy, capture whether policy makers devote their attention to the debate about the capacity of self-rule of the regions. This includes all bills, oral questions or direct references of the Prime Minister related to the shift of issue jurisdiction, and the problems associated to this process (subcategory 1); and the debate about fiscal autonomy, like bills, to modify the allocation of fiscal resources across territories; or oral questions that highlight the non-fulfillment of previous agreements, the inequalities among territories associated to the distribution of fiscal resources, or the special status of some regions, mainly the Basque country and Navarra (subcategory 2).

In order to capture the capacity of regions to participate in the policy-making process at the national level (shared rule), we have created two additional subcategories. The third subcategory summarizes the demands for institutional reform of the Senate, the Spanish Constitution, reform of the Estatutos de Autonomía, or judicial bodies like the Constitutional Court. This dimension also considers policy positions about the participation of regions in EU matters and other international organizations like the Unesco, for example (recognition of the role of regions in foreign affairs); demands to alter the participation of regions in executive bodies of decision-making; and demands to modify the representation of the Spanish government in the territory of regions (like the elimination of the office of regional governor (*Gobernadores Civiles*). The fourth subcategory refers to debates about the capacity of regional governments (and regional parties with representation in the Spanish Parliament) to impose a veto on the agenda of the Spanish executive, or vice versa, and/or the opinion of the Prime Minister members of the Spanish government about the activities of regional governments. This subcategory may be important in the case of oral questions, and to some extent speeches, having little importance in the case of bills. Finally, the fifth subcategory does not relate to the shared rule/self-rule distinction but refers to the symbolic arena, including oral questions about the use of official languages, cultural heritage, recognition of regional sport federations, etc.

Appendix D: List of EU Events Coded

The following events where coded as major EU-related events, in the months indicated. See Chapter 7 for an explanation of how we use these codes (Table 7.2 in particular).

Signature of EU Treaties and referendums: Single European Act (February 1986), Maastricht (February, June, September, December 1992; May 1993; April 1995; September 2000; September 2003); Amsterdam (October 1997, May 1998); Nice (February and June 2001; October 2002); European Constitution (October, November and December 2004;

February, April, May, June, July 2005; February 2006); Lisbon (December 2007, June 2008, October 2009).

Signature of accession treaties: Spain and Portugal (June 1985); Austria, Finland, Sweden (June 1994), Malta Slovenia, Hungary, Lithuania, Slovakia, Poland, Cyprus, Czech Republic, Estonia, Latvia (April 2003), Romania, Bulgaria (April 2005), Croatia (December 2013).

Signature of EU-related treaties and agreements: Schengen Agreement (June 1985); Prüm Convention (May 2005); Treaty Establishing the European Stability Mechanism (February 2012); European Fiscal Compact (March 2012); Treaty on Stability Coordination and Governance in the Economic and Monetary Union (March 2012).

EMU-related events and convergence reports: creation of the European Monetary Institute (January 1994), creation of the European Central Bank (December 1998); Stability and Growth Pack-related events (July 1997, July 1998, January 1999, March 2005, September 2010, March 2011); creation of the euro and introduction of notes and coins (January 1999 and January 2002); approval of rescue programs for Greece, Ireland, Portugal, Spain, and Cyprus (May 2010, February 2010, November 2010, April 2011, May 2011, July 2012, June 2012); publication of convergence reports (December 1993, November 1996, March 1998, May 2000, May 2002; October 2004; May and December 2006, May 2007, May 2008, May 2010, May 2012, June 2013).

Council meetings: June and December 1986; June and December 1987; February, June, and December 1988; June, November, and December 1989; June, October, and December 1990; April, June, and December 1991; June, October, and December 1992; June, October, and December 1993; June, July, and December 1994; June, October, and December 1995; March, June, October, and December 1996; May, June, November, and December 1997; May, June, October, and December 1998; February, March, April, June, October, and December 1999; March, June, October, and December 2000, March, June, September, October, and December 2001; March, June, October, and December 2002; February, March, April, June, October, and December 2003; March, June, November, and December 2004; March, June, October, and December 2005; March, June, October, and December 2006; March, June, October, and December 2007; March, June, July, September, October, November, and December 2008; March, April, June, September, October, November, and December 2009; February, March, May, June, September, October, and December 2010; February, March, June, July, October, and December, 2011; January,

March, May, June, October, November, and December 2012; February, March, May, June, and December 2013.

Spanish Council Presidencies: January-June 1989; July-December 1995; January-June 2002; January-June 2010.

European Parliament elections: June 1984, 1987, 1989, 1994, 1999, 2004, 2009, 2014.

Changes in the Presidency of the European Commission: November 1993; January 1995; March 1999; September 1999; November 2004 and 2009.

Notes

1 Policy Dynamics in Democratic Spain

1. See the OCDE or the Economist Intelligence Unit annual report among others. Also the World Bank good governance indicators are available at http://info.worldbank.org/governance/wgi/.

2 Data and Methods

1. Spanish Policy Agendas Project, available here: www.ub.edu/spanishpolicy-agendas.
2. A quasi-sentence is a unit of analysis that contains a single message or statement so that a natural sentence can contain one or more quasi-sentences.
3. Spanish Parliament webpage, available at: www.congreso.es.
4. Comparative Agendas Project (CAP), available here: www.comparativea-gendas.info/.
5. Substantive agendas are tangible, deal with concrete policy measures, and are mostly designed to the passage of formal legislation (Pritchard, 1992; Pritchard and Berkowitz, 1993; Walgrave and Van Aelst, 2006).
6. As defined in Article 99 of the Spanish Constitution, the proposed candidate (designated by the King after consultation with the representatives designated by the parliamentary group with representation in the Spanish Parliament) has to submit to the Parliament the political program of the government he intends to form.
7. In all other CAP projects, political activities related to political decentralization are coded as subtopic 2001 that includes all types of intergovernmental relations, including local governments. This description is especially relevant for those researchers willing to compare Spain with any other country included in the CAP project.
8. The regional laws dataset covers the period 1980–2008 in the case of Catalonia and the Basque Country, and 1982–2008 in the cases of Andalusia and Galicia.
9. The number of polls conducted varies across years. See www.cis.es for detailed methodological information.
10. The Bureau is responsible for the management of the Chamber and is vested with the collegial representation of it. It consists of the Chamber President, four Deputy Speakers, and four Secretaries. Parliament shall elect the members of the Bureau in its constituent sitting. In the election of the Speaker, each member shall write a single name on the ballot paper. The candidate having obtained the votes of the overall majority of members of the House shall be elected. If no candidate obtains such a majority on the first ballot, the election shall be repeated among the two candidates who have achieved the highest number of votes, and the candidate who obtains more votes shall

then be elected. The four Deputy Speakers shall be elected simultaneously. Each member shall write a single name on the ballot paper. The four candidates obtaining the highest number of votes shall be elected in that order. The four Secretaries shall be elected in the same manner (Chamber rules, art. 37).

11. Spokespersons of parliamentary groups make up the Board of Spokesmen. Meetings of the Board are chaired by the Speaker of Parliament (Chamber rules, art. 39).
12. Delegation can occur through an act of basic principles (*Ley de Bases*) that contains authorization to pass a text in sections (*texto articulado*), or through an ordinary law designed to consolidate several legal statutes into one (*texto refundido*).
13. Executive bills are initiated within a ministerial department, and once passed in the Council of Ministries they go directly to the Parliament where after publication in the *Boletín Oficial de la Cortes* they initiate the formal legislative procedure.
14. Shannon's H Entropy = – Σ p(xi)*logn p(xi) where xi represents a dimension, p(xi) is the proportion of total attention the dimension receives, and logn p(xi) is the log of the proportion of attention the dimension receives, using the total number of possible dimensions as the base of the log (Boydstun et al., 2014).
15. To calculate media attention to different topics we have considered only those stories with political content for the period 1996–2011.

3 Broken Promises

1. http://eur-lex.europa.eu/legal-content/AUTO/?uri=CELEX:31983L0129&qid=1403776738007&rid=4
2. The process took place with a too-long delay, but despite the resistance of many Cajas, in early 2010, 27 Cajas (out of 45) were already involved in merge processes. A large set of reforms were developed including tightening rules on how banks value their portfolios (like acquired properties), and increasing transparency on their balance sheets (mainly developed to avoid any possibility of hiding bad loans) (see Royo, 2013, for a description).
3. For example, this means from 2010, elected officials could not be members of the governing board of saving institutions (with some exceptions like La Caixa, where about 25 percent of the governing board were politicians) (see Royo, 2013, for a description).
4. According to Article 135.3, limits on public debt will be determined by the reference value established by the treaty on the functioning of the EU, currently at 60 percent of GDP. Indirectly, debt dynamics will also be constrained by a limit on the general government's structural budget deficit (Article 135.2). All these agreements are further developed by organic law, which in any case cannot surpass the 3 percent of the GDP, with some margin of variation depending on external circumstances.
5. All data is available at the IGAE (Intervención General de la Administración del Estado, Gobierno de España) and also the OCDE economic indicators.

6. See http://observatoridepoliticaexterioreuropea.wordpress.com/espana-en-europa/
7. The alliance of civilizations is an initiative proposed by José Luis Rodríguez Zapatero at the 59th General Assembly of the United Nations in 2005. It was co-sponsored by Turkish Prime Minister Recep Tayyip Erdoğan. The goal of the initiative was to defuse tensions between the Western and Islamic worlds. To read more about the initiative see www.unaoc.org.
8. The *Gobernadores Civiles* were created in the nineteenth century and for more than a century have been a symbol of political oppression. Formally, they were the institutional representatives of the Spanish government in each of the provinces in Spain. Among other things, they were in charge of the police. The Constitution did not mention this institution. Despite this, it took more than two decades to eliminate the *Gobernadores Civiles*, which occurred in 1997 (*Ley Orgánica 6/1997, de Organización y Funcionamiento de la Administración General del Estado*) (see Baena, 1997).
9. Actually, some regional governments already developed specific relations on this issue. In Catalonia, Miquel Caminal Badia was the first director of the *Memorial Democràtic*, created in 2007.
10. One of the most visible cases of this failure is the assassination of Yoyes in the hands of ETA in 2006, after leaving the terrorist group. There is a vast literature about terrorism in Spain (see Sanchez Cuenca, 2001; Muro, 2008; Llera, 1993, among others).
11. There is only one short episode of conversations between the PP and ETA in 1998 (the first unilateral truce of ETA). For about 14 months, the PP government negotiated with ETA, arriving to some important agreements like the reallocation of 100 ETA prisoners to prisons located in the Basque Country.
12. The *"Acuerdo por las libertades y Contra el terrorismo"* is a document of 10 articles signed by the PP and PSOE, December 8, 2000. This document is available at http://www.interior.gob.es/.
13. Newspapers widely covered this declaration of Rajoy. See, for example, Rajoy: "Usted traiciona a los muertos y ha revigorizado a una ETA moribunda," *El País*, May 12, 2005.
14. The *Ley Orgánica 7/1985 de 1 de Julio sobre derechos y libertades de los extranjeros en España* (so-called Ley de Extranjería) had three main goals: (1) to respond to one of the requirements related to joining the EEC (actually to incorporate mechanisms to deal with immigration were of little practical relevance in Spain, given the very small immigration flows in the 1980s); (2) to eliminate the existing regulatory dispersion; and (3) to define the goals of immigration policy according to European demands, primarily border control.
15. *Ley Orgánica 4/2000 de los Derechos y Libertades de los Extranjeros y su Integración Social* (Organic Law about the rights and freedom of foreigners and their social integration).
16. *Ley Orgánica 8/2000* and *Ley Orgánica 11/2003 de medidas concretas en material de seguridad ciudadana, violencia doméstica e integración social de los extranjeros.*
17. *Ley Orgánica 2/2009 de 11 de diciembre, de reforma de la Ley Orgánica 4/2000, de 11 de enero, sobre derechos y libertades de los extranjeros en España y su integración social.*

18. See, for example, *"Determinantes de la brecha salarial de género en España", Consejo Superior de Cámaras en Colaboración con el Ministerio de Sanidad, Servicios Sociales e Igualdad.*
19. Along the first term, the PP also concluded the process of political decentralization and transfer of issue jurisdiction of health and education issues to those CCAA that followed the slowtrack to become CCAA. See Chapter 6 for further details.
20. Family policy was defined by the *Plan Integral de Apoyo a la Familia* (Integral plan to support families), and *Ley 39/1999 para promover la conciliación de la vida familiar y laboral de las personas trabajadoras.*
21. Families, and especially women, assume the responsibility of taking care of the young (education from 0 to 3 is noncompulsory and the number of public nurseries does not cover existing demand); the aged (pensions, especially widow pensions, are lower and public residences do not cover existing demand either, forcing the elderly to rely on their adult children); and the unemployed (more than 25 percent of Spaniards under 30 live with their parents, a pattern that is particularly strong among those who are unemployed or not fully secure in their positions).
22. Parliamentary rules guarantee these privileges, without major modifications from 1978 to present. First, parliamentary rules establish all governmental bills should be taken into consideration and discussed in full in the *Congreso de los Diputados*. That is, in contrast to other advanced democracies, the Spanish Parliament do not have the capacity to veto executive bills a priori (Döring, 1995, 2001). Second, all MPs can introduce amendments to executive bills. However, most of the time the governing party has the majority of seats in the Spanish Parliament, or sets stable governmental agreements with small and regional parties (see Table 2.2), the adoption rate (percentage of bills passed into law) is very high. Finally, the party in government also has special prerogatives to pass laws in case of urgent need, the so-called decree-laws (see Chapter 2). All decree-laws have to be ratified within 30 days by the *Congreso de los Diputados*, and in contrast to executive bills, MPs cannot introduce any amendments to their content – the voting is a yes or no to the overall content of the decree-law – (Guerrero, 2005). This is a major difference in relation to executive bills, which in more than 75 percent of the cases are passed with at least one amendment.
23. Note that in some cases, parliamentary groups may decide to introduce a bill dealing with the content of a decree-law. The purpose of this initiative is to guarantee the discussion about issues that are considered of special relevance. Still, this may end up being a symbolic procedure without major political consequences.
24. Note that final voting of laws is only possible for the case of organic laws and decree-laws, so these results only refer to these two legislative instruments (see Table A3.1 in the appendix for a full technical explanation).

4 The Content of the Government-Opposition Game

1. If we take into consideration a poll from the Centro de Investigaciones Sociológicas (www.cis.es) conducted in 2009.

2. *Resolución de la Presidencia del Congreso de los Diputados, de 18 de Junio de 1996, sobre el Desarrollo del artículo 188 del Reglamento, relativo a pregunta para respuesta oral en Pleno; Resolución de la Presidencia del Congreso de los Diputados, de 10 de junio de 2008, sobre desarrollo del artículo 188 del Reglamento, relativo a preguntas para respuesta oral en Pleno (modificada por Resolución de la Presidencia, de 28 de diciembre de 2011).*
3. Note that all calculations are made from 1986 mainly because it is only from 1986 when the board of the Congreso de los Diputados introduced some rationality in terms of number of sessions, questions per group, etc. Also, before 1986 the Far Left do not have a parliamentary group.
4. Note that here the period under study is 1986 to 2011, and thus an interaction term controlling the impact of changing rules depending on whether a party is governing or giving support to the government is not applicable (the PSOE is governing from 2008 to 2011).
5. Public opinion data comes from a single poll conducted by the CIS (www.cis.es) in April-May 2009 (studio 2.799), which contains two explicit questions about issue ownership for nine issues (question 25: Would you evaluate from 0 to 10, the performance of the PSOE government in the following areas: immigration (subtopic 230 and 529), education (topic 6), terrorism (subtopic 1260), social policy (topic 3 and 13), economy (topic 1 and 15), control of crime (topic 12 except 1204), political decentralization (subtopic 2070), justice (subtopic 1204), environment (topic 7 and 21), and foreign affairs (topic 19); and question 26: do you think that the PP would be doing a better job, worse or the same than the PSOE government for the same areas).

5 Media and Politics in Spain

1. For more information, see the annual reports of the *Asociación para la Investigación de Medios de Comunicación* (http://www.aimc.es/)
2. *Ley 46/1983 Reguladora del Tercer Canal.*
3. *Ley 10/1988 de Television Privada.*
4. *Ley 41/1995 de Television Local.*
5. *Ley 12/1997 de 24 abril de Liberalización de las Comunicaciones.*
6. *Ley 7/2010 General de la Comunicación audiovisual.*
7. Data on media parallelism is only available in terms of readership. The CIS has introduced a specific question in several polls designed to capture this link between voters of political parties and readers, and ideological positions and readers of newspapers. Here, we provide data from two different surveys: studio no. 2920/0 Post-electoral elecciones generales 2011, question 25a; and studio no. 2798 Barómetro abril 2009, question 13c. In both questions respondents were asked about which newspaper they read to get information about politics. Here, the ideological scale goes from 1 (Far Left) to 10 (Far Right).
8. Media audience has increased modestly in the last decades, and increases are mainly linked to digital newspapers (mostly free to the general public) and the consolidation of free newspapers. Despite that most readers still prefer the print version of newspapers (about 80 percent), in 2012 the number of readers of digital newspapers was more than 20 percent (AMC, various years).

9. "The Freedom of the Press Index" assesses the degree of print, broadcast, and Internet freedom in every country in the world, analyzing the events and developments of each calendar year. Ratings are determined through an examination of three broad categories: the legal environment in which the media operate, political influences on reporting and access to information, and economic pressures on content and the dissemination of news (Freedom of the Press, 2013: 3). Detailed information about the methodology and rating review process is explained in any of the annual reports and the web page of Freedom House (www.freedomhouse.org).
10. See http://freedomhouse.org/report/freedom-press-2014/methodology#.U9uKg-N_t8F

6 The Politics of Decentralization

1. Actually, from 1981 to 1983 the Constitutional Court sentenced 54 regional conflicts (*conflictos y recursos de inconstitucionalidad*), and from 1984 to 1990 more than 220 (see Aja, 2014, for a detailed analysis).
2. Distribution of competences in Spain follows a quite complicated model. Article 149 of the Spanish Constitution lists 32 areas of legislative competence in which the center has exclusive competence. Article 148 lists the issues that each region may assume responsibilities for. Many of these competences are shared with the central government. The central government has exclusive jurisdiction over state functions like defense and foreign affairs, citizenship, immigration, and civil rights; and sets and implements most policies related to economic stabilization (mostly Europeanized after the Economic and Monetary Union), transportation, energy, and communications. For issues with shared jurisdiction, the central government sets the basic principles, from which regional governments develop their own laws, setting their own goals and implementation strategies in their territories. This applies to welfare issues like health, education, labor, social policy in general, environmental issues, and taxation, with important differences between the Basque Country, Navarra, and the rest of CCAA.
3. The fiscal reform was regulated by *Ley Orgánica 3/1996, de 27 de diciembre, de modificación parcial de la Ley Orgánica 8/1980, de 22 de septiembre, de financiación de las Comunidades Autónomas and Ley 14/1996 de 30 de diciembre, de cesión de tributos del Estado a las Comunidades Autónomas y de medidas fiscales complementarias*, later defined by *Ley 31/1997, de 4 de agosto, de modificación del régimen de cesión de tributos del Estado a la Generalidad de Cataluña y de fijación del alcance y condiciones de dicha cesión. Ley 21/2001, de 27 de diciembre, por la que se regulan las medidas fiscales y administrativas del nuevo sistema de financiación de las Comunidades Autónomas de régimen común y Ciudades con Estatuto de Autonomía. Ley 17/2002, de 1 de julio, del régimen de cesión de tributos del Estado a la Generalidad de Cataluña y de fijación del alcance y condiciones de dicha cesión*.
4. These surveys are available at the CIS web page (www.cis.es): survey number 2286 (*Instituciones y Autonomías 1998*), survey number 2610 (*Barómetro Autonómico I 2005*), survey number 2829 (*Barómetro Autonómico II 2010*), survey number 2956 (*Barométro Autonómico III*).

5. Data comes from the CIS (www.cis.es). This figure summarizes the responses of Spanish citizens of 18 years or more about the preferences between different alternatives to the territorial organization of the state. The survey covers all the territory of Spain including Ceuta and Melilla, and has been conducted regularly from 1984 to present. The precise question is: *"Le voy a presentar ahora algunas fórmulas alternativas de organización territorial del Estado en España. Dígame, por favor, con cuál está Ud. más de acuerdo"* ("I am going to present different alternatives about the territorial organization of the state. Please, inform me about which is the alternative you prefer"). From 1984 to April 2009 respondents had to choose among four different answers, while from 2009 on, respondents have five different options. This is described in the figure. The survey numbers are 1441/0 24, 1495/0 24, 1558/0 27, 1715/0 24, 1764/0 21, 1773/0 31, 1851/0 18, 2123/0 20, 2154/0 11, 2211/0 25, 2228/0 27, 2286/0 14, 2309/0 37, 2455/0 15, 2535/0 8, 2610/0 12, 2667/0 8, 2736/0 38, 2757/0 12,2799/0 39, 2828/0 24, 2834/0 25, 2843/0 20, 2847/0 31, 2859/0 27, 2920/0 19, 2930/0 31, 2951/0 20, 2954/0 30, 2960/0 26, 2966/0 24, 2972/0 22, 2976/0 21, 2978/0 28, 2981/0 29, 2984/0 22, 2987/0 27, 2993/0 21, 2997/0 31, 3001/0 25, 3005/0 25, 3008/0 25, 3011/0 20, 3013/0 23.
6. The decisions taken by the CARUE only have a binding character when they refer to issues for which CCAA have exclusive jurisdiction, and when decisions are taken by unanimity. This constrains the possibilities of Catalonia, or any other regions, to influence on the strategy of the central government on EU matters.
7. More information at: http://www.seap.minhap.gob.es/ca/areas/politica_autonomica/participacion-ccaa-eu/ccaa_y_ue/Participacion_CCAA_Consejo_Ministros.html
8. CiU, PNV, and BNG signed a common agreement (the *Declaració de Barcelona*) in 1998 designed to transform the Estado de las Autonomías into a confederation formed by the historical regions (Catalonia, Galicia, and the Basque Country) and the rest of Spain.

7 Europeanization

1. García-Lupato (2008) analyzes the role of Europe in parliamentary competition considering government investiture speeches and budget policy debates.
2. We considered as Europeanized those laws partially or totally defined by an EU-binding regulatory act (see Palau and Chaqués-Bonafont, 2012). In the case of oral questions, they are considered as having EU content if they refer to the EU or mention any policy or EU institution in the title of the question.
3. From 1977 to 2007, 264 questions were introduced in the EAC, which contrasts with the 2299 questions introduced in the Justice and Internal Affairs Committee or the 1423 introduced in the Education Committee.
4. This Committee was established in 1986 (*Comisión mixta Congreso-Senado para las Comunidades Europeas*) to facilitate and to promote the Parliament's involvement in EU issues, and to formalize the systematic and permanent transfer of information from government to Parliament to avoid the duplication of information in the Congress, Senate, and various parliamentary committees (Jones, 2000). Originally it included 15 members – nine from

the Congress and six from the Senate – designed according to representation criteria, and its functions were mainly limited to access EU-related information and to control legislative decrees passed to implement EU law. One of the main limitations of this committee was that the government could inform it "a posteriori," so most debates concerned issues on which decisions had already been taken by EU institutions (Storini, 2005). As a result, different reforms were introduced during the 1980s and 1990s to guarantee the effective representation of all parliamentary groups and also to accommodate the demands of the Maastricht treaty that urged member states to guarantee that national parliaments receive legislative proposals from the Commission at an earlier stage in the process. The committee was reformed again after the Lisbon treaty to incorporate the protocol on the principle of subsidiarity and proportionality, and to facilitate the participation of government's representatives in the EAC before the celebration of Council meetings.

5. In Denmark, the EAC has the power to limit governmental bargaining discretion by specifying a negotiation mandate – a collective decision, which must be supported by a majority of MPs (Finke and Melzer, 2012). Similarly, in the UK, there is "scrutiny reserve" requirement, giving the Parliament the right to demand time from the government to be able to consider matters and debate them before the government may vote in the Council of Ministers (Holzhacker, 2005: 435).
6. The correlation between agenda capacity (number of questions asked) and number of oral questions asked with EU content is 0.419** in the case of the PP, 0.348** for the PSOE, 0.166* IU, 0.361** CiU, and 0.333** PNV.
7. The Eurobarometer series "benefits perceived from European integration" is only available until 2011.

8 Framing the Abortion Debate

1. Baumgartner et al. (2008) studied a similar issue in the context of the United States: the death penalty. As long as it is seen from a moral or religious perspective, it is hard to envision rapid shifts in public opinion. But they demonstrate that the issue in fact was reframed away from the abstract questions of morality to very practical questions of whether the government can always be trusted to execute only the guilty. The rise of the "innocence frame" had a powerful effect on the debate, driving public opinion down, even though most Americans may not have shifted their views on the abstract theoretical question of whether the penalty of death is ever justified.
2. The law also defines different periods for each exception. In the case of rape, abortion must take place within the first 12 weeks. In the case the fetus is severely handicapped abortion must take place within the first 22 weeks. No time restriction applied in case of serious harm to the mother. The capacity to decide whether or not these exceptions apply never relies on women but physicians that must elaborate a report explaining the health status of the mother, and/or certify the health status of the fetus.
3. The PSOE 1982 electoral manifesto included the reform of the abortion in certain circumstances.

4. This bill was designed to include a fourth circumstance where women would be allowed to legally access abortions: when the pregnancy represented a personal, family or social conflict provided the woman had received enough information about the social and economic assistance they would have in case of having the child, and if it was practiced during the first twelve weeks of pregnancy.

9 Transformations in Spanish Politics, 1982–2013

1. Instituto Nacional de Estadística, www.ine.es

References

Adler, S. and J. D. Wilkerson (2012) *Congress and the Politics of Problem Solving* (Cambridge: Cambridge University Press).

Aguilar, P. (1996) *Memoria y olvido de la Guerra Civil española* (Madrid: Alianza Editorial).

Aguilar. S. (1997) *El reto del medio ambiente: conflictos e intereses en la politica medio-ambiental europea* (Madrid: Alianza Editorial).

Aja, E. (2012) *Inmigración y Democracia* (Madrid: Alianza Editorial).

Aja, E. (2014) *El Estado Autonómico y Reforma Federal* (Madrid: Alianza Editorial).

Alba, C. (2001) "Bureaucratics Politics in Spain: A Long-Lasting Tradition" in B. Peters and J. Pierre, *Politicians, Bureaucrats and Administrative Reform* (London: Routledge/Ecpr Studies in European Political Science).

Alcántara, M. and A. Martínez (1997) *Elecciones Autonómicas en España 1980–1997* (Madrid: Centro de Investigaciones Sociológicas).

Anderson, C. J. (1998) "When in Doubt, Use Proxies: Attitudes toward Domestic Politics and Support for European Integration", *Comparative Political Studies*, 31: 569–601.

Anderson, C. D. (2006) "Economic Voting and Multilevel Governance: A Comparative Individual-Level Analysis", *American Journal of Political Science*, 50(2): 449–463.

Arregui, J. (2007) "Europeización y Transformación Institucional en el Sistema Político Español" in F. Morata and G. Mateo (eds) *España en Europa, Europa en España (1986–2006)* (Barcelona: Fundació CIDOB).

Arrojo, P. (2005) *"El Reto Ético de la Nueva Cultura del Agua"*. Editorial Paidós. Barcelona.

Backes and P. Moreau (eds) (2012) *The Extreme Right in Europe:* Current Trends and Perspectives (Göttingen: Vandenhoeck & Ruprecht).

Baena, M. (1997) *Curso de Ciencia de la Administración* (Madrid: Tecnos).

Barahona De Brito, A., C. Gonzalez Enriquez and P. Aguilar (2001): *The Politics of Memory, Transitional Justice in Democratizing Societies* (New York, Oxford: Oxford Studies in Democratization).

Barbé, E. (1999) *La Política Europea de España*. (Barcelona: Ariel).

Barreiro, B. (1998) "Judicial Review and Political Empowerment: Abortion in Spain", *West European Politics*, 21(4): 147–162.

Barreiro, B. (2000) *Democracia y conflicto moral: la política del aborto en España e Italia* (Madrid: Istmo).

Bassols, R. (1995) *España en Europa. Historia de la adhesión a la CE, 1957–1985* (Madrid: Estudios de Política Exterior).

Baumgartner, F. R. and L. Chaqués-Bonafont (2015) "All News Is Bad News: Newspaper Coverage of Political Parties in Spain", *Political Communication*, 1–24.

Baumgartner, F. R., S. L. De Boef and A. E. Boydstun (2008) *The Decline of the Death Penalty and the Discovery of Innocence* (New York: Cambridge University Press).

Baumgartner, F. R. and B. D. Jones (2009 [1993]) *Agendas and Instability in American Politics* (Chicago: University of Chicago Press).

Baumgartner, F. R. and B. D. Jones (2015) *The Politics of Information: Problem Definition and the Course of Public Policy in America* (Chicago: University of Chicago Press).

Baumgartner, F. R., B. D. Jones and J. Wilkerson (2011) "The Dynamics of Policy Change in Comparative Perspective", *Comparative Political Studies*, 44(8): 947–972.

Bel, G. (2011) *Espanya, capital París. Tots els camins porten a Madrid* (Barcelona: La Campana).

Bel, G. and A. Costas (2001) "La privatización y sus motivaciones en España: de instrumento a política", *Revista de Historia Industrial*, 19–20: 105–132.

Bellamy, R. and D. Castiglione (2013) "Three Models of Democracy, Political Community and Representation in the EU", *Journal of European Public Policy*, 20(2).

Bennett, W. L. (1990) "Toward a Theory of Press-State Relations in the United States", *Journal of Communication*, 40(2): 103–125.

Bergés, L. (2010) "Poder politico, económico y comunicativo en la sociedad neoliberal", *Revista Latina de Comunicación Social*, 65: 244–254.

Bertelli, A. M. and P. John (2013) "Public Policy Investment: Risk and Return in British Politics", *British Journal of Political Science*, 43 (4), 741–773. doi:10.1017/S0007123412000567.

Bevan, S. and W. Jennings (2014) "Representation, Agendas and Institutions", *European Journal of Political Research*, 53(1): 37–56.

Bianculli, A. and J. Jordana (2013) "The Unattainable Politics of Child Benefits Policy in Spain", *Journal of European Social Policy*, 23: 504–520.

Boix, C. (1998) *Political Parties, Growth and Equality: Conservative and Social Democratic Economic Strategies in the World Economy* (Cambridge: Cambridge University Press).

Boix, C. (2006) "The Roots of Democracy", *Policy Review*, 35 (February–March), http://www.hoover.org/research/roots-democracy.

Bonal, X. (1998) "La política educativa: dimensiones de un proceso de transformación (1976–1996)", in R. Gomá and J. Subirats (eds) *Políticas Públicas en España: Contenidos, Redes de Actores y Niveles de Gobierno*, (153–175) (Barcelona: Ariel).

Börzel, T. (2000) "Why There Is No 'Southern Problem': On Environmental Leaders and Laggards in the European Union", *Journal of European Public Policy*, 7(1): 141–162.

Bosch, N., M. Espasa and A. Solé (eds) (2010) *The political economy of inter-regional fiscal flows*. (Cheltenham: Edward Elgar).

Boydstun, A. E. (2013) *Making the News: Politics, the Media and Agenda Setting* (Chicago: University of Chicago Press).

Boydstun, A. E., S. Bevan and H. F. Thomas III. (2014) "The Importance of Attention Diversity and How to Measure It", *Policy Studies Journal*, 42(2): 173–196.

Brasslof, A. (1998) *Religion and Politics in Spain: The Spanish Church in Transition, 1962–96* (Basingstoke: Palgrave Macmillan).

Brambor, T., W. R. Clark and M. Golder. (2006) "Understanding interaction models: Improving empirical analyses". *Political analysis*, 14(1): 63–82.

Brouard, S., O. Costa and T. Köning (2012) *The Europeanization of Domestic Legislatures: The Empirical Implications of the Delors' Myth in Nine Countries, Studies in Public Choice* (New York: Springer).

Brouard, S., F. Gougou, I. Guinaudeau and S. Persico (2014) "Un effet de campagne: le déclin de l'opposition des Français au nucléaire en 2011–2012", *Revue Française de Science Politique*, 63(6): 1051–1079.

Brouard, S. and I. Guinaudeau (2014) "Policy Beyond Politics? Party Politics, Public Opinion and the French Pro-Nuclear Energy Policy", *Journal of Public Policy*, available online http://CJO2014. doi:10.1017/S0143814X14000221.

Budge, I. and D. J. Farlie (1983). *Explaining and Predicting Elections: Issue Effects and Party Strategies in Twenty-Three Democracies*. (London: Allen & Unwin).

Budge, I. and R. I. Hofferbert (1990) "Mandates and Policy Outputs: U.S. Party Platforms and Federal Expenditures", *American Political Science Review*, 84: 111–132.

Budge, I. and H. Keman. (1990) *Parties and Democracy* (New York: Oxford).

Budge, I., H. Klingemann, A. Volkens, J. Bara, E. Tanenbaum, C. R. Fording, D. Hearl, H. M. Kim, M. McDonald and S. Mendez (2001) *Mapping Policy Preferences. Estimates for Parties, Electors, and Governments 1945–1998* (Oxford: Oxford University Press).

Caminal, M. and F. Requejo (2010) *Federalisme i plurinacionalitat. Teoria i Anàlisi de Casos* (Barcelona: IEA).

Campa, J. M. and M. F. Guillén (1999) "The Internalization of Exports: Firm- and Location-Specific Factors in a Middle-Income Country", *Management Science*, 45(11): 1463–1479.

Capo, J. (1994) "Oposición y minorías en las legislaturas socialistas", *Revista Española de Investigaciones Sociológicas* (REIS), 66: 91–113.

Capo, J. (2003) "The Spanish Parliament in a Triangular Relationship, 1982–2000", *Journal of Legislative Studies*, 9(2): 107–129.

Castells, M. (2009) *Communication Power* (Oxford: Oxford University Press).

Chaqués-Bonafont, L. (2002) *Estructura y Política farmacéutica/* (Madrid: Centro de Investigaciones Sociológicas).

Chaqués-Bonafont, L. (2004) *Redes de Políticas Públicas* (Madrid: Centro de Investigaciones Sociológicas).

Chaqués-Bonafont, L. and F. R. Baumgartner (2013) "Newspaper Attention and Policy Activities in Spain", *Journal of Public Policy*, 33: 65–88.

Chaqués-Bonafont, L. and A. M. Palau (2009) "The Dynamics of Policy Change: A Comparative Analysis of the Food Safety and Pharmaceutical Policy in Spain", *Journal of Public Policy*, 29:1, 103–126.

Chaqués-Bonafont, L. and A. M. Palau (2011a) " Comparing Law-Making Activities in a Quasi-Federal System of Government: The Case of Spain", *Comparative Political Studies*, 44(8): 1089–1119.

Chaqués-Bonafont, L. and A. M. Palau (2011) "Assessing the Responsiveness of Spanish Policymakers to the Priorities of their Citizens", *West European Politics*, 34(4): 706–730.

Chaqués-Bonafont, L. and A. M. Palau (2012) "From Prohibition to Permissiveness: A Two Wave Change on Morality Issues in Spain" in I. Engeli, C. Green-Pedersen, L. T. Larsen (eds) *Morality Politics in Western Europe* (London: Palgrave Macmillan), 62–87.

Chaqués-Bonafont, L., A. M. Palau and L. Muñoz (2014) "Policy Promises and Governmental Activities in Spain" in C. Green-Pedersen and S. Walgrave (eds)

Tracing Political Attention: A Novel Approach to Comparative Politics (Chicago: Chicago University Press).

Chryssochoou, D. N., S. Stavridis and M. J. Tsinisizelis (1998) "European Democracy, Parliamentary Decline and the 'Democratic Deficit' of the European Union", *Journal of Legislative Studies*, 4(3): 109–129.

Cienfuegos, M. (2001) "Las Cortes Españolas ante la Integración Europea" in C. Closa (ed.) *La Europeización del sistema politico español* (Madrid: Istmo).

Closa, C. (1995) "Spain: The Cortes and the EU: Growing Together", *Journal of Legislative Studies,* 1(3): 1136–1150.

Closa, C. and P. Heywood (2004) *Spain and the European Union* (New York: Palgrave Macmillan).

Colino, C. and del Pino, E. (2014) "Spanish Federalism in Times of Crisis: Towards Further Centralization and Growing Tensions?", in P. Peterson and D. Nadler (eds) *Fiscal Crisis and the Centralization of Political Power* (Washington: Brookings Institution Press).

Curran, J., S. Iyengar, A. B. Lund and I. Salovaara-Moring (2009) "Media System, Public Knowledge and Democracy: A Comparative Study", *European Journal of Communication*, 24(1): 5–26.

De la Calle, L. and I. Sánchez-Cuenca (2012) "Rebels Without a Territory: An Analysis of Non-territorial Conflicts in the World, 1970–1997", *Journal of Conflict Resolution*, 56(4): 580–603. Online first: April 18, 2012.

Del Aguila, R. (1982) "La transición a la democracia en España: reforma, ruptura y consenso", *Revista de Estudios Políticos* (Nueva Época), 25: 101–127.

Della Porta, D. and A. Mattoni (2014) *Spreading Protest: Social Movements in Times of Crisis* (London: ECPR).

Del Pino, E. (2013) "The Spanish Welfare State from Zapatero to Rajoy: Recalibration to Retrenchment" in A. Botti and B. N. Field (eds) *Politics and Society in Contemporary Spain: From Zapatero to Rajoy* (New York: Palgrave Macmillan).

De Wilde, P. (2011) "No polity for old politics? A framework for analyzing politicization of European integration", *Journal of European Integration*, 33(5): 559–575.

Díez Medrano, J. (1995) *Actitudes de los Españoles ante la Unión Europea* (Madrid: CIS).

Díez Medrano, J. (2003) *Framing Europe: Attitudes to European Integration in Germany, Spain, and the United Kingdom* (Princeton, NJ: Princeton University Press).

Díez Medrano, J. (2007) "La opinión pública española y la integración europea" in F. Morata and G. Mateo (eds) *España en Europa-Europa en España (1986–2006)* (Barcelona: Fundació CIDOB).

Döring, H. (1995) "Parliaments and Majority Rule in Western Europe", *New York: St. Martin's Press.*

Downs, A. (1957) *An Economic Theory of Democracy* (New York: Harper and Row).

Dunphy, R. (2004) *Contesting Capitalism? Left Parties and European Integration* (Manchester: Manchester University Press).

Engeli, I., C. Green-Pedersen and L. T. Larsen (eds) (2012) *Morality Politics in Western Europe: Parties, Agendas and Policy Choices* (London: Palgrave Macmillan).

Esping-Andersen, G. (2000) *Fundamentos sociales de las Economías Post-industriales* (Barcelona: Ariel).

Esping-Andersen, G. (2009) *The Incomplete Revolution: Adapting to Women's New Roles* (Cambridge: Polity Press).

Fernández-Albertos, J. and D. Manzano. (2012) "The Lack of Partisan Conflict over the Welfare State in Spain", *South European Society and Politics,* 17(3): 427–447.

Finke, D. and M. Melzer (2012) *Parliamentary Scrutiny of EU Law proposals in Denmark – Why do Governments request a Negotiation Mandate?* HIS Political Science Series, 127, https://www.ihs.ac.at/publications/pol/pw_127.pdf. Wien: Institute for Advanced Studies.

Finke, D. and Dannwolf, T. (2013) "Domestic Scrutiny of European Union Politics: Between Whistle Blowing and Opposition Control", *European Journal of Political Research,* 52: 715–746.

Fishman, R. M. (2012) "Anomalies of Spain's Economy and Economic Policy-making", *Contributions to Political Economy,* 31: 67–76.

Follesdal, A. and S. Hix (2006) "Why there is a democratic deficit in the EU: A response to Majone and Moravcsik", *Journal of Common Market Studies,* 44(3): 533–562.

Fraile, M. (2010) "Widening or Reducing the Knowledge Gap? Testing the Media Effects on Political Knowledge in Spain (2004–2006)", *International Journal of Press/Politics,* 16(2): 163–184.

Franklin, M. and P. Norton (eds) (1993) *Parliamentary Questions* (Oxford: Clarendon Press.).

Freedom House report. (2013) https://freedomhouse.org/report-types/freedom-press#.VPiXrPmG_HU

Font, J. and Rodríguez, E. (2006) "The Spanish Referendum on the EU Constitution: Issues, Party Cues and Second Order Effects". Paper presented at the 64th annual meeting of the MidWest Political Science Association, Chicago, 20–23 April.

Garcia-Lupato, F. (2008) "*Europeanisation and Political Parties: Saliency and Usage of Europe in Italian and Spanish Electoral and Parliamentary Competition*", PhD thesis, University of Sienna.

Gentzkow, M. and J. Shapiro (2006) "Media Bias and Reputation", *Journal of Political Economy,* 114(2): 280–316.

Graber, D. A. (2003) *The Power of Communication: Managing Information in Public Organizations.* (Washington: CQ Press).

Green-Pedersen, C. (2006) "Long-Term Changes in Danish Party Politics: The Rise and Importance of Issue Competition", *Scandinavian Political Studies,* 29(3): 219–235.

Green-Pedersen, C. (2007) "The growing importance of issue competition: The changing nature of party competition in Western Europe", *Political Studies,* 55(4): 608–628.

Green-Pedersen, C. (2010) "Bringing Parties into Parliament the Development of Parliamentary Activities in Western Europe", *Party Politics,* 16(3): 347–369.

Green-Pedersen, C. (2012) "A Giant Fast Asleep? Party Incentives and the Politicisation of European Integration", *Political Studies,* 60: 115–130.

Green-Pedersen, C. and P. B. Mortensen (2010) "Who Sets the Agenda and Who Responds to It in the Danish Parliament?", *European Journal of Political Research,* 49(2): 257–281.

Green-Pedersen, C. and R. Stubager (2010) "The Political Conditionality of Mass Media Influence: When Do Parties Follow Mass Media Attention?" *British Journal of Political Science,* 40: 633–677.

Green-Pedersen, C. and S. Walgrave (2014): *Agenda Setting, Policies and Political Systems* (Chicago: Chicago University Press).

Guibernau, M. (2013) *Belonging: Solidarity and Division in Modern Societies* (Cambridge: Polity Press).
Guillén, A. M. and León, M. (eds) (2011) *The Spanish Welfare State in European Context* (London: Ashgate).
Guillén, M. F. (2005) *The Rise of Spanish Multinationals: European Business in the Global Economy* (Cambridge: Cambridge University Press).
Guillén, M. F. and E. Garcia-Canal (2010) *The New Multinationals: Spanish Firms in a Global Context* (Cambridge: Cambridge University Press).
Guinaudeau, I. and A. M. Palau (2013) "The impact of exogenous and endogenous factors on the media coverage of European integration: a comparative analysis of the UK, France and Spain (1990–2013)", Paper presented at the ECPR General Conference, Bordeaux, September 4–7, 2013.
Gunther, R. (ed.) (1993) *Politics, Society and Democracy: The Case of Spain* (Boulder, CO: Westview Press).
Gunther, R. and J. R. Montero (2009) *The Politics of Spain* (Cambridge: Cambridge University Press).
Gunther, R., J. R. Montero and J. Botella (2004) *Democracy in Modern Spain* (New Haven, CT: Yale University Press).
Gunther, R. and A. Mughan (2000) *Democracy and the Media: A Comparative Perspective* (Cambridge: Cambridge University Press).
Hall, P. A. (1993) "Policy Paradigms, Social Learning, and the State: The Case of Economic Policymaking in Britain." *Comparative Politics*, 25: 275–296.
Hallin, D. C. and P. Mancini (2004) *Comparing Media Systems: Three Models of Media and Politics* (New York: Cambridge University Press).
Heywood, P. (1998) "Power Diffusion or Concentration? In Search of the Spanish Policy Process", *Western European Politics*, 21(4): 103–123.
Hix, S. (1999) *The Political System of the European Union* (New York: Palgrave Macmillan).
Holzhacker, R. L. (2002) "National Parliamentary Scrutiny over EU Issues. Comparing the Goals and Methods of Governing and Opposition Parties", *European Union Politics*, 3(4): 459–479.
Hooghe, L. and G. Marks (2001) *Multi-Level Governance and European Integration* (Boulder, CO: Rowman & Littlefield).
Hooghe, L. and G. Marks (2009) "Postfunctionalist Theory of European Integration: From Permissive Consensus to Constraining Dissensus", *British Journal of Political Science*, 39(1): 1–23.
Hooghe, L., G. Marks and A. H. Schakel (2008) "Regional Authority in 42 Democracies, 1950–2006. A Measure and Five Hypotheses", *Regional and Federal Studies*, 18(2–3): 111–302.
Hooghe, L., G. Marks and A. H. Schakel (2010) *The Rise of Regional Authority: A Comparative Study of 42 Democracies (1950–2006)* (London: Routledge).
Iyengar, S. (1991) *Is Anyone Responsible? How Television Frames Political Issues* (Chicago: University of Chicago Press).
Iyengar, S. and J. A. McGrady (2007) *Media Politics: A Citizen's Guide.* (New York: W.W. Norton & Company).
Jachtenfuchs, M. (2001) "The governance approach to European integration", *Journal of Common Market Studies* 39(2): 245–264.
Jennings, W., S. Bevan, A. Timmermans, G. Breeman, S. Brouard, L. Chaqués-Bonafont, C. Green-Pedersen, P. John, P. B. Mortensen and A. M. Palau (2011)

"Effects of the Core Functions of Government on the Diversity of Executive Agendas", *Comparative Political Studies*, 44(8): 973–1000.

John, P., A. Bertelli, W. Jennings and S. Bevan (2014) *Policy Agendas in British Politics* (Basingstoke: Palgrave Macmillan).

John, P., S. Bevan and W. Jennings (2011) "The Policy-Opinion Link and Institutional Change: The Legislative Agenda of the United Kingdom and Scottish Parliaments", *Journal of European Public Policy*, 18(7): 1052–1068.

Jones, B. D. (2001) *Politics and the Architecture of Choice* (Chicago: University of Chicago Press).

Jones, B. D. and F. R. Baumgartner (2005) *The Politics of Attention* (Chicago: University of Chicago Press).

Jones, B. D., H. Larsen-Price and J. Wilkerson (2009) "Representation and American Governing Institutions", *The Journal of Politics*, 71(1): 277–290.

Jones, E. D. (2007) "Grupos mediáticos y culturales en España", *Zer*, 22: 183–214

Jones, E. (2009) "Output legitimacy and the global financial crisis: Perceptions matter", *Journal of Common Market Studies*, 47(5): 1085–1105.

Jones, R. (2000) Beyond the Spanish State: Central Government, Domestic Actors and the EU. (HoundMills: Palgrave).

Jordana, J. (2014) "Multiple Crises and Policy Dismantling in Spain: Political Strategies and Distributive Implications", *Political Studies Review*, 12(2): 224–238.

Keating, M. (1998) *The New Regionalism in Western Europe: Territorial Restructuring and Political Change* (Northampton: Edward Elgar Publishing).

Keating, M. (2001) *Nations against the State: The New Politics of Nationalism in Quebec, Catalonia, and Scotland* (Basingstoke: Palgrave Macmillan).

Kingdon, J. W. (1984) *Agendas, Alternatives and Public Policies* (Boston: Little Brown).

Klingemann, H D., R. I. Hofferbert and I. Budge (1994) *Parties, Policies and Democracy* (Boulder: Westview Press).

Kriesi, H., E. Grande, R. Lachat, M. Dolezal, S. Bornschier and T. Frey (2008) *West European Politics in the Age of Globalization* (Cambridge: Cambridge University Press).

Krugman, P. (2012) *Acabad ya con esta crisis* (Barcelona: Crítica).

Lago, I. and S. Lago (2011) *Descentralización y control electoral de los gobiernos en España* (Barcelona: Institut d'Estudis Autonòmics).

Lijphart, A. (1999) *Patterns of Democracy: Government Forms and Performance in Thirty-Six Countries* (New Haven, CT: Yale University Press).

Linz, J. J. and J. R. Montero (eds) (1986) *Crisis y cambio: electores y partidos en la España de los años ochenta* (Madrid: Centro de Estudios Constitucionales).

Llera, F.J. (1993) "Conflicto en Euskadi Revisited." In: R. Gunther (ed.) *Politics, Society, and Democracy: The Case of Spain*. (Boulder, CO: Westview Press).

Lombardo, E. (2003) "La europeización de la política española de igualdad de género", *Revista Española de Ciencia Política*, 9: 63–80.

López-Nieto, L. (2001) "Las Cortes Generales", in M. Alcántara and A. Martínez (eds) *Política y Gobierno en España* (215–242) (Valencia: Tirant lo Blanch).

Lord, C. and Pollak, J. (2010) "Representation and Accountability: Communicating Tubes", *West European Politics*, 33(5): 968–988.

Magone, J. M. (2004) *Contemporary Spanish Politics* (London: Routledge).

Mair, P. (1998) Party System Change: Approaches and Interpretations (Oxford: The Clarendon Press).

Mair, P., W. C. Müller and F. Plasser (eds) (2004) *Political Parties and Electoral Change: Party Responses to Electoral Markets* (London: Sage).
Manin, B., A. Przeworski and S. C. Stokes (1999) "Elections and Representation" in A. Przeworski, S. C. Stokes and B. Manin (eds) *Democracy, Accountability and Representation* (29–54) (Cambridge: Cambridge University Press).
Maravall, J. M. (1997) *Regimes, Politics, and Markets* (Oxford: Oxford University Press).
Maravall, J. M. (1999) "Accountability and Manipulation" in A. Przeworski, S. C. Stokes and B. Manin (eds) *Democracy, Accountability and Representation* (154–196) (Cambridge: Cambridge University Press).
Maravall, J. M. (2013) *Las promesas políticas* (Barcelona: Galaxia Gutenberg/ Círculo de Lectores).
Maravall, J. M. and I. Sánchez Cuenca (2008) *Controlling Governments: Voters, Institutions, and Accountability* (Cambridge: Cambridge University Press).
Martin, S. (2011) "Parliamentary questions, the behaviour of legislators, and the function of legislatures: An introduction", *Journal of Legislative Studies*, 17(3): 259–270.
Martínez, A. (ed.) (2000) *El Congreso de los diputados en España: funciones y rendimiento* (Madrid: Tecnos).
Matia, F. J. (1999) *Parlamentos Nacionales y Derecho Comunitari* (Madrid: Centro de Estudios Políticos y Constitucionales).
Maurer, A. and W. Wessels (eds) (2001) *National Parliaments on their Ways to Europe. Losers or Latecomers?* (Baden Baden: Nomos Verlag).
McChesney, R. (2003) "Theses on Media Deregulation", *Media, Culture and Society*, 25: 125–133.
McCombs, M. E. and D. L. Shaw (1972) "The agenda setting function of mass media", *Public Opinion Quarterly*, 36(2): 176–187.
McCombs, M. and J. Zhu (1995) "Capacity, Diversity, and Volatility of the Public Agenda: Trends from 1954 to 1994", *Public Opinion Quarterly*, 59: 495–525.
Molina, I. (2001) "La adaptación a la Unión Europea del poder ejecutivo español" in C. Closa (ed.) *La europeización del sistema político español* (Madrid: Istmo).
Montero, J. R., J. Font and M. Torcal (eds) (2006) *Ciudadanos, asociaciones y participación en España* (Madrid: Centro de Investigaciones Sociológicas).
Montero, J. R. and M. Torcal (1991) "Política y cambio cultural en España: una nota sobre la dimensión postmaterialista", Ponencia presentada en el Simposio sobre Mutación del sistema de valores en las sociedades europeas y magrebíes, Barcelona.
Morales, L. and M. Giuni (2011) *Social Capital, Political Participation and Migration in Europe. Migration, Minorities and Citizenship* (Basingstoke: Palgrave Macmillan).
Morata, F. (ed.) (2004) Gobernanza multinivel en la Unión Europea, Tirant lo Blanch, Valencia.
Morata, F. (1998) *La Unión Europea. Procesos, actores y políticas* (Barcelona: Ariel).
Morata, F. and G. Mateo (2007) *España en Europa-Europa en España (1986–2006)* (Barcelona: Fundació CIDOB).
Moravsick, A. (1994) "Why the European Union Strengthens the State: Domestic Politics and International Cooperation", *Center for European Studies, Working Paper Series* 52.
Moreno, L. (2009) "Modelo social europeo y políticas sociales: una evaluación formativa institucional", *Gestión y Análisis de Políticas Públicas*, 2: 11–32.

Mújica, A. and I. Sánchez-Cuenca (2006) "Consensus and Parliamentary Opposition", *Government and Opposition*, 41(1): 86–108.

Muñoz, J. (2012) *La construcción política de la identidad española: ¿del nacional-catolicismo al patriotismo democrático?* (Madrid: Centro de Investigaciones Sociológicas).

Muñoz Márquez, L. (2013) "Treinta años de cooperación al desarrollo en España: análisis de la continuidad y cambio en las políticas públicas", *Revista Mexicana de Análisis Político y Administración Pública*, 2(2): 77–105.

Muro, D. (2008) *Ethnicity and Violence: The Case of Radical Basque Nationalism* (London: Routledge/Cañada Blanch Series in Contemporary Spain).

Navarro, V. (2009) *El subdesarrollo social en España: Causas y Consecuencias* (Madrid: Diario Público).

Newton, K. and M. Brynin. (2001) "The National Press and Party Voting in the UK", *Political Studies*, 49(2): 265–285.

Nieto, A. (1997) *Corrupción en la España democrática* (Barcelona: Ariel)

Norris, P. (2000) *A Virtuous Circle* (Cambridge: Cambridge University Press).

Norris, P. (2001) *Digital Divide? Civic Engagement, Information Poverty and the Internet in Democratic Societies* (Cambridge: Cambridge University Press).

Olmeda, J. A., S. Parrado and C. Colino (2012) *Las Administraciones Públicas en España* (Valencia: Tirant lo Blanch).

Ortega, E. and J. Peñalosa (2012) "The Spanish ecomomic crisis: key factors and growth challenges in the Euro area", *Documentos Ocasiones*, Banco de España, 1201.

Palau, A. M. (2009) "Continuidad y cambio en la política de seguridad alimentaria española", *Revista Española de Ciencia Política*, 21: 47–68.

Palau, A. M. (2013) "El impacto de la Unión Europea en la agenda legislativa estatal y autonómica (1986–2007)", *Revista de Estudios Políticos*, 160: 69–102.

Palau, A. M. and L. Chaqués-Bonafont (2012) "Europeanization of legislative activity in Spain" in S. Brouard, O. Costa and T. Köning (eds) *The Europeanization of domestic legislatures. The empirical implications of the Delors' Myth in Nine Countries* (New York: Springer).

Palau, A. M. and F. Davesa (2013) "El impacto de la cobertura mediática de la corrupción en la opinión pública española (1996–2009)", *Revista Española de Investigaciones Sociológicas*, 144: 97–125.

Palau, A. M., L. Muñoz and L. Chaqués-Bonafont (2014) "Government opposition dynamics in Spain under the pressure of economic collapse and debt crisis", *Journal of Legislative Studies*, published online, http://www.tandfonline.com/doi/abs/10.1080/13572334.2014.939565#.VDuiVrDkdqE

Pasquino, G. (1995). "Executive-Legislative Relations" in R. Gunther, N. Diamandouros and H. J. Puhle (eds) *The Politics of Democratic Consolidation: Southern Europe in Comparative Perspective* (Baltimore: Johns Hopkins University Press), 261–283.

Pérez-Díaz, V. (1996) *España puesta a prueba 1976–1996*, (Madrid: Alianza Editorial).

Pérez-Díaz, V. (2001) "La familia española ante la educación de sus hijos", *Colección Estudios Sociales* 5, Fundación la Caixa.

Pérez-Díaz, V., J. Mezo and B. Á. Miranda (1996) *Política y Economía del Agua en España*. (Madrid: Circulo de Empresarios).

Peters, G. and J. Piercre (2013) *Multi-level Governance and Democracy: A Faustian Bargain* (Oxford: Oxford University Press).

Petrocik, J. (1996) "Issue-Ownership in Presidential Elections with a 1980 Case Study", *American Journal of Political Science*, 40(3): 825–850.
Petrocik, J. R., W. L. Benoit and G. J. Hansen (2003). "Issue Ownership and Presidential Campaigning, 1952–2000," *Political Science Quarterly*, 118: 599–626.
Piedrafita, S., F. Steinberg and J. I. Torreblanca (2006) *20 años de España en la Unión Europea 1986–2006* (Madrid: Real Instituto Elcano).
Prindham, G. and J. M. Magone (2006) "The environment, socio-economic transformation and political change in the New Southern Europe", in R. Gunther and N. Diamandouros (eds) *The Chanching Functions of the State in the New Southern Europe*. (Baltimore, Oxford: Johns Hopkins University Press, Oxford University Press), 262–304.
Przeworski, A. (1991) *Democracy and the Market* (Cambridge: Cambridge University Press).
Przeworski, A., S. C. Stokes and B. Manin (1999) *Democracy, Accountability, and Representation* (Cambridge: Cambridge University Press).
Pugglisi, R. and J. M. Snyder Jr. (2008) "Media Coverage of Political Scandals", *NBER Working Papers* 14598, National Bureau of Economic Research, Inc.
Quaglia, L. and S. Royo (2014) "Banks and the political economy of the sovereign debt crisis in Italy and Spain", *Review of International Political Economy*, http://dx.doi.org/10.1080/09692290.2013.877059.
Quintanilla, M. (2000): "Los partidos políticos españoles ante el proceso de integración europea", Revista de Estudios Políticos, 108, abril-junio 2000.
Ramon, R. (2007) "España y la Política Agrícola Común" in F. Morata and G. Mateo (eds) *España en Europa-Europa en España (1986–2006)* (Barcelona: CIDOB Fundació CIDOB).
Raunio, T. (1999) "Always one step behind? National Legislatures and the European Union", *Government and Opposition*, 11(3–4): 319–342.
Raunio, T. (2005) "Holding Governments Accountable in European Affairs: Explaining Cross-National Variation". *Journal of Legislative Studies*, 11(3–4), 319–342.
Reinares, F. (2004) *"El Nuevo Terrorismo Islamista. Del 11-S al 11-M"*, (Madrid: Temas de Hoy).
Requejo, F. and C. Miquel (2010) *Political Liberalism and Plurinational Democracies* (London: Routledge).
Requena, M. (2005) "The secularization of Spanish Society, changing religious practice", *South European Society and Politics*, 10(3): 369–390.
Richardson, J. J. (ed.) (2006) *European Union: Power and Policy-Making* (Abingdon: Routledge).
Rico, A. and R. Sabés (2000) *Health Care Systems in Transition* (Copenhagen: WHO Regional Office for Europe, European Observatory on Health Care System).
Riker, W. H. (1986) *The art of political manipulation* (New Haven: Yale University Press).
Rodríguez-Aguilera, C. (2014) "Is Spanish Untied Left a Eurospectic party? An analysis of its European manifestos of 2004 and 2009", Paper presented at MPSA Chicago Annual Conference.
Rodríguez-Aguilera, C. (2013) *Euroscepticism, Europhobia and Eurocriticism. The Radical Parties of the Right and Left vis-à-vis the European Union* (Brussels: PIE-Peter Lang).
Rodríguez Aguilera, C. and F. Morata (2008) "Europa en la refriega política Española (y en especial, catalana): los partidos políticos ante el Tratado Constitucional",

in H. Jörg Trenz, A. J. Menéndez and F. Losada (eds) *¿Y por fin somos europeos? La comunicación política en el debate constituyente europeo* (Madrid: Aguilera Daykinson).
Royo, S. (2013) "How did the Spanish financial system survive the first stage of the global crisis?", *Governance*, 26: 631–656.
Rubio Llorente, F. (1996). "El constitucionalismo de los Estados integrados en Europa", *Revista Española de Derecho Constitucional*, (48): 9–33.
Russo, F. and M. Wiberg (2010) "Parliamentary questioning in 17 European parliaments: some steps towards comparison", *Journal of Legislative Studies*, 16(2): 215–232.
Saalfeld, T. (2000). "Members of parliament and governments in Western Europe: Agency relations and problems of oversight", *European Journal of Political Research*, 37(3): 353–376.
Salido, O. (2011) "Female employment and work-lefe balance policies in Spain", in A. M. Guillén and M. León (ed.) *The Spanish Welfare State in European Context* (London: Ashgate).
Sampedro, V. (2004) "Identidades mediáticas. La lógica del régimen de visibilidad contemporánea", *Sphera Pública, Revista de Ciencias Sociales y de la Comunicación*, (4): 17–35.
Sampedro, V. and F. S. Pérez (2008) "The 2008 Spanish General Elections: 'Antagonistic Bipolarization' Geared by Presidential Debates, Partisanship, and Media Interest", *International Journal of Press/Politics*, 13(3): 336–344.
Sánchez, I. (1999) "El déficit democrático en la Unión Europea", in I. Llamazares and F. Reinares (eds) *Aspectos políticos y sociales de la integración europea*, 94–116 (Valencia: Tirant lo Blanch).
Sánchez Cuenca, I. (2000) "The political basis of support for European integration", *European Union Politics*, 1(2): 141–171.
Sánchez Cuenca, I. (2001) *ETA contra el Estado: Las estrategias del terrorismo* (Barcelona: Tusquets).
Sánchez de Dios, M. and M. Wiberg (2011) "Questioning in European Parliaments", *Journal of Legislative Studies*, 17(3): 354–367.
Scharpf, F. W. (1999) *Governing in Europe. Effective and Democratic?* (Oxford: Oxford University Press).
Schattschneider, E. E. (1975[1960]). *The Semisovereign People: A Realist's View of Democracy in America* (Fort Worth, TX: Harcourt Brace Jovanovich).
Schmidt, V. A. (2006) *Democracy in Europe: The EU and National Polities* (Oxford: Oxford University Press).
Schmitter, P. C. (2004) "Neo-neo-functionalism", in A. Weiner and T. Diez (eds) *European Integration Theory*, 45–74 (Oxford: Oxford University Press).
Seymour-Ure, C. (1974) *The Political Impact of Mass Media* (London: Constable/ Sage).
Seymour-Ure, C. (2003) *Prime Ministers and the Media: Issues of Power and Control* (Oxford: Blackwell).
Simon, H. A. (1947) *Administrative Behavior: A Study of Decision Making Processes in Administrative Organizations* (New York: Palgrave Macmillan).
Simon, H. A. (1971) "Designing Organizations for an Information-Rich World", in M. Greenberger *Computers, Communication, and the Public Interest* (Baltimore, MD: The Johns Hopkins Press) 37–72.

Soroka, S. N. and C. Wlezien (2010) *Degrees of Democracy: Politics, Public Opinion and Policy*. (Cambridge: Cambridge University Press).

Stiglitz, J. (2012) *The Price of Inequality How today's Divided Society Endangers Our Future* (New York: Norton & Company).

Stimson, J. A. (1991) *Public Opinion in America: Moods, Cycles, and Swings* (Boulder, CO: Westview).

Stimson, J. A., M. B. Mackuen and R. S. Erikson (1995) "Dynamic Representation", *American Political Science Review*, 89(3): 543–565.

Stimson, J., C. Thiébaut and V. Tiberj (2012) "The Evolution of Policy Attitudes in France", *European, Union Politics*, 13(2): 293–316.

Stimson, J., V. Tiberj and C. Thiébaut (2013) "The Evolution of Political Attitudes and Policy Preferences in France" in A. Cole, S. Meunier and V. Tiberj (eds) *Developments in French Politics 5* (Basingstroke: Palgrave Macmillan), 170–185.

Stokes, S. (1999) "Political Parties and Democracy", *Annual Review of Political Science*, (2): 243–267.

Stokes, S. (2001) *Mandates and Democracy* (Cambridge: Cambridge University Press).

Storini, C. (2005) *Parlamentos Nacionales y Unión Europea* (Valencia: Tirant lo Blanch).

Strömback, J. and P. Van Aelst (2013) "Why Political Parties adapt to the Media: Exploring the Fourth Dimension of Mediatization", *International Communication Gazette*, 75(4): 341–358.

Subirats, J., Q. Brugue and R. Goma (2001) "Multilevel governance and Europeanization" in K. Featherstone and G. Kazamias (eds) *Europeanization and Souther Periphery*, 95–118. (London: Frank Cass).

Subirats, J. and R. Gallego (2002) Veinte años de autonomías en España. Leyes, políticas públicas, instituciones y opinión pública (Madrid: Centro de Investigaciones Sociológicas).

Szczerbiak, A. and P. Taggart (2003) "Theorising Party-based Euroscepticism: Problems of Definition, Measurement and Causality", paper presented at the in European Union Studies Association International Conference, Nashville, Tennessee, March 27–29, 2003.

Szmolka, I. (1999) "Opiniones y actitudes de los españoles ante el proceso de Integración Europea", *Opiniones y Actitudes*, 21 (Madrid: Centro de Investigaciones Sociológicas).

Szmolka, I. (2007) Veinte años de pertenencia de España a la Unión Europea: actitudes de los españoles ante el proceso de integración comunitaria (Madrid: Centro de Investigaciones Sociológicas).

Talbert, J. C. and M. Potoski (2002) "Setting the legislative agenda: the dimensional structure of bill cosponsoring and floor voting", *Journal of Politics*, 64(3): 864–891.

Thesen, G. (2013) "When Good News is Scarce and Bad News is Good: Government Responsibilities and Opposition Possibilities in Political Agenda Setting", *European Journal of Political Research*, 52: 364–389.

Thompson, J. B. (2001) *El escándalo político. Poder y visibilidad en la era de los medios de comunicación* (Barcelona: Paidós).

Topaloff, L. K. (2012) *Political Parties and Euroscepticism* (New York: Palgrave Macmillan).

Torcal, M. and L. Medina (2002) "Ideología y voto en España 1979–2000: los procesos de reconstrucción racional de la identificación ideológica", *Revista Española de Ciencia Política*, 6: 57–96.

Torreblanca, J. I. (2001) *The Reuniting of Europe: Promises, Negotiations and Compromises* (Aldershot /Burlington: Ashgate).

Torreblanca, J. I. (2005) "¿Adiós a los fondos? Las claves para entender la posición de España a la hora de negociar el presupuesto de la UE para 2007–2013", *Real Instituto Elcano, Documento de Trabajo 21/2005*.

Vallespín, F. (2012). *La mentira os hará libres: realidad y ficción en la democracia* (Barcelona: Galaxia Gutenberg).

Van Dalen, A. and P. Van Aelst (2014) "The media as political agenda-setter in comparative perspective: Journalists perceptions of media power in eight West-European countries", *West European Politics*, 37(1): 42–64.

Varone, F., C. R. Allison and E. Montpetit (2006) "Regulating Biomedicine in Europe and North America: A Qualitative Comparative Analysis", *European Journal of Political Research*, 45(2): 317–343.

Vilanova, P. (2006) *Orden y desorden a escala global* (Barcelona: Síntesis).

Vintró, J. (2006) La investidura parlamentaria del gobierno: perspectiva comparada y Constitución Española (Madrid: Congreso de los Diputados).

Vliegenthart, R. and S. Walgrave (2011) "Content matters: The dynamics of parliamentary questioning in Belgium and Denmark", *Comparative Political Studies*, 1031–1059.

Walgrave, S. and M. Nuytemans (2009) "Party manifesto change: friction or smooth adaptation? A comparative analysis in 25 countries (1945–1998)", *American Journal of Political Science*, 53: 190–206.

Walgrave, S., S. Soroka and M. Nuytemans (2007) "The Mass Media's Political Agenda-Setting Power: A Longitudinal Analysis of Media, Parliament, and Government in Belgium (1993 to 2000)", *Comparative Political Studies*, 41(6): 814–836.

Walgrave, S. and P. Van Aelst (2006) "The Contingency of the Mass Media's Political Agenda Setting Power: Towards A Preliminary Theory", *Journal of Communication*, 56: 88–109.

Walgrave, S., F. Varone and P. Dumont (2006) "Policy with or without parties? A comparative analysis of policy priorities and policy change in Belgium (1991 to 2000)", *Journal of European Public Policy*, 13(7): 1021–1038.

Wiberg, M. (1995) "Parliamentary questioning: control by communication?", in H. Döring (ed.) *Parliaments and Majority Rule in Western Europe* (Frankfurt/New York: Campus Verlag), 179–222.

Wolfe, M., B. D. Jones and F. R. Baumgartner (2013) "A Failure to Communicate: Agenda Setting in Media and Policy Studies, Political Communication", 30(2): 175–192.

Wolfsfeld, G. and T. Sheafer (2006) "Competing Actors and the Construction of Political News: The Contest over Waves in Israel", *Political Communication*, 23(3): 333–354.

Wonka, A. and Rittberger, B. (2014): "The Ties that Bind? Intra-party Information Exchanges of German MPs in EU Multi-level Politics", *West European Politics*, 37:(3), 624–643.

Zubek, R. (2011) "Core Executives and the Co-ordination of EU Law Transposition: Evidence from New Member States" *Public Administration*, 89(2): 433–450.

Index

CPSIA information can be obtained at www.ICGtesting.com
Printed in the USA
LVOW03*2346180815

450611LV00012B/326/P

9 781137 328786